LONDON MATHEMATICAL SOCIETY LECTURE NOTE SERIES

Managing Editor: Professor J.W.S. Cassels, Department of Pure Mathematics and Mathematical Statistics, University of Cambridge, 16 Mill Lane, Cambridge CB2 1SB, England

The books in the series listed below are available from booksellers, or, in case of difficulty, from Cambridge University Press.

London Mathematical Society Lecture Note Series. 174

Lectures on Mechanics

Jerrold E. Marsden
Department of Mathematics and EECS, University of California at Berkeley

CAMBRIDGE
UNIVERSITY PRESS

Cambridge

New York   Port Chester   Melbourne   Sydney

Published by the Press Syndicate of the University of Cambridge
The Pitt Building, Trumpington Street, Cambridge CB2 1RP
40 West 20th Street, New York, NY 10011-4211, USA
10 Stamford Road, Oakleigh, Victoria 3166, Australia

First published 1992

Printed in Great Britain at the University Press, Cambridge

Library of Congress cataloguing in publication data available

British Library cataloguing in publication data available

ISBN   0 521 42844 0

Contents

Preface

Many of the greatest mathematicians — Euler, Gauss, Lagrange, Riemann, Poincaré, Hilbert, Birkhoff, Atiyah, Arnold, Smale — were well versed in mechanics and many of the greatest advances in mathematics use ideas from mechanics in a fundamental way. Why is it no longer taught as a basic subject to mathematicians? *Anonymous*

I venture to hope that my lectures may interest engineers, physicists, and astronomers as well as mathematicians. If one may accuse mathematicians as a class of ignoring the mathematical problems of the modern physics and astronomy, one may, with no less justice perhaps, accuse physicists and astronomers of ignoring departments of the pure mathematics which have reached a high degree of development and are fitted to render valuable service to physics and astronomy. It is the great need of the present in mathematical science that the pure science and those departments of physical science in which it finds its most important applications should again be brought into the intimate association which proved so fruitful in the work of Lagrange and Gauss. *Felix Klein, 1896*

These lectures cover a selection of topics from recent developments in the geometric approach to mechanics and its applications. In particular, we emphasize methods based on symmetry, especially the action of Lie groups, both continuous and discrete, and their associated Noether conserved quantities veiwed in the geometric context of momentum maps. In this setting, *relative equilibria*, the analogue of fixed points for systems without symmetry are especially interesting. In general, relative equilibria are dynamic orbits that are also group orbits. For the rotation group $SO(3)$, these are uniformly rotating states or, in other words, *dynamical motions in steady rotation*.

Some of the main points to be treated are as follows:

- The stability of relative equilibria analyzed using the method of separation of internal and rotational modes, also referred to as the block diagonalization or normal form technique.

- Geometric phases, including the phases of Berry and Hannay, are studied using the technique of reduction and reconstruction.

- Mechanical integrators, such as numerical schemes that exactly preserve the symplectic structure, energy, or the momentum map.

- Stabilization and control using methods especially adapted to mechanical systems.

- Bifurcation of relative equilibria in mechanical systems, dealing with the appearance of new relative equilibria and their symmetry breaking as parameters are varied, and with the development of complex (chaotic) dynamical motions.

A unifying theme for many of these aspects is provided by reduction theory and the associated mechanical connection for mechanical systems with symmetry. When one does reduction, one sets the corresponding conserved quantity (the momentum map) equal to a constant, and quotients by the subgroup of the symmetry group that leaves this set invariant. One arrives at the reduced symplectic manifold that itself is often a bundle that carrying a connection. This connection is induced by a basic ingredient in the theory, the *mechanical connection* on configuration space. This point of view is sometimes called the gauge theory approach to mechanics. The geometry of reduction and the mechanical connection is an important ingredient in the decomposition into internal and rotational modes in the block diagonalization method, a powerful method for analyzing the stability and bifurcation of relative equilibria. The holonomy of the connection on the reduction bundle gives geometric phases. When stability of a relative equilibrium is lost, one can get bifurcation, *solution symmetry breaking*, instability and chaos. The notion of *system symmetry breaking* in which not the solutions but the equations loose symmetry, is also important but here is treated only by means of some simple examples.

Two related topics that are discussed are control and mechanical integrators. One would like to be able to control the geometric phases with the aim of, for example, controlling the attitude of a rigid body with internal rotors. With mechanical integrators one is interested in designing numerical integrators that exactly preserve the conserved momentum (say angular momentum) and either the energy or symplectic structure, for the purpose of accurate long time integration of mechanical systems. Such integrators are becoming popular methods as their performance gets tested in specific applications. We include a chapter on this topic that is meant to be a basic introduction to the theory, but not the practice of these algorithms.

This work proceeds at a reasonably advanced level but has the corresponding advantage of a shorter length. For a more detailed exposition of many of these topics suitable for beginning students in the subject, see my book with Tudor Ratiu, *Mechanics and Symmetry* (Springer-Verlag, 1992).

The work of many of my colleagues from around the world is drawn upon in these lectures and is hereby gratefully acknowledged. In this regard, I especially thank Vladimir Arnold, Judy Arms, John Ball, Tony Bloch, David Chillingworth, Richard Cushman, Michael Dellnitz, Arthur Fischer, Mark Gotay, Marty Golubitsky, John Harnad, Darryl Holm, Phil Holmes, John Guckenheimer, Jacques Hurtubise, Vivien Kirk, P.S. Krishnaprasad, Debbie Lewis, Robert Littlejohn, Ian Melbourne, Vincent Moncrief, Richard Montgomery, George Patrick, Tom Posbergh, Tudor Ratiu, Alexi Reyman, Gloria Sanchez de Alvarez, Shankar Sastry, Jürgen Scheurle, Juan Simo, Ian Stewart, Greg Walsh, Steve Wan, Alan Weinstein, Shmuel Weissman, Steve Wiggins, and Brett Zombro. The work of others is cited at appropriate points in the text.

I would like to especially thank David Chillingworth for organizing the LMS lecture series in Southampton, April 15–19, 1991 that acted as a major stimulus for preparing the written version of these notes. I would like to also thank the Mathematical Sciences Research Institute and especially Alan Weinstein and Tudor Ratiu at Berkeley for arranging a preliminary set of lectures along these lines in April, 1989, and Francis Clarke at the Centre de Recherches Mathématique in Montréal for his hospitality during the Aisenstadt lectures in the fall of 1989. Thanks are also due to Phil Holmes and John Guckenheimer at Cornell, the Mathematical Sciences Institute, and to David Sattinger and Peter Olver at the University of Minnesota, and the Institute for Mathematics and its Applications, where several of these talks were given in various forms. I also thank the Humboldt Stiftung of Germany, Jürgen Scheurle and Klaus Kirchgässner who provided the opportunity and resources needed to put the lectures to paper during a pleasant and fruitful stay in Hamburg during the first half of 1991. I also acknowledge a variety of research support from NSF, DOE, and AFOSR that helped make the work possible. I thank several participants of the lecture series and other colleagues for their useful comments and corrections. I especially thank Hans Peter Kruse, Oliver O'Reilly, Rick Wicklin, Brett Zombro and Florence Lin in this respect.

Very special thanks go to Barbara for typesetting the lectures and for her support in so many ways. Thomas the Cat also deserves thanks for his help with our understanding of 180° cat manouvers. This work was not responsible for his unfortunate fall from the roof (resulting in a broken paw), but his feat did prove that cats can execute 90° attitude control as well.

Chapter 1

Introduction

This chapter gives an overview of some of the topics that will be covered so the reader can get a coherent picture of the types of problems and associated mathematical structures that will be developed.[1]

1.1   The Classical Water Molecule and the Ozone Molecule

An example that will be used to illustrate various concepts throughout these lectures is the *classical (non-quantum) rotating "water molecule"*. This system consists of three particles interacting by interparticle conservative forces (one can think of springs connecting the particles, for example) so that the total energy of the system, which will be taken as our Hamiltonian, is the sum of the kinetic and potenial energies. This system is shown in Figure 1.1.1. The interesting special case of three equal masses gives the "ozone" molecule.

We use the term "water molecule" mainly for terminological convenience. The full problem is of course the classical *three body problem* in space. However, thinking of it as a rotating system evokes certain constructions that we wish to illustrate.

Imagine this mechanical system rotating in space and, simultaneously, undergoing vibratory, or *internal* motions. We can ask a number of questions:

- How does one set up the *equations of motion* for this system?

---

[1] We are grateful to Oliver O'Reilly, Rick Wicklin, and Brett Zombro for providing a helpful draft of the notes for an early version of this lecture.

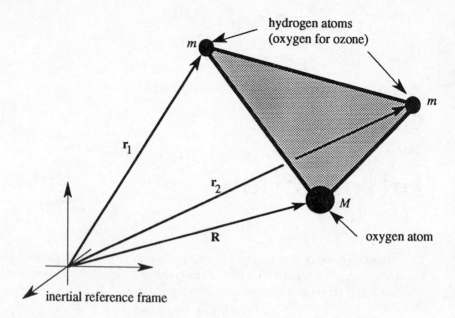

Figure 1.1.1: The rotating and vibrating water molecule.

- Is there a convenient way to describe *steady rotations*? Which of these is stable? When do bifurcations occur?

- Is there a way to separate the *rotational* from the *internal motions*?

- How do vibrations affect overall rotations? Can one use them to *control* overall rotations? To *stabilize* otherwise unstable motions?

- Can one separate symmetric (the two hydrogen atoms moving as mirror images) and non-symmetric vibrations using a discrete symmetry?

- Does a deeper understanding of the classical mechanics of the water molecule help with the corresponding quantum problem?

It is interesting that despite the old age of classical mechanics, new and deep insights are coming to light by combining the rich heritage of knowledge already well founded by masters like Newton, Euler, Lagrange, Jacobi, Laplace, Riemann and Poincaré, with the newer techniques of geometry and qualitative analysis of people like Arnold and Smale. I hope that already the classical water molecule and related systems will convey some of the spirit of modern research in geometric mechanics.

The water molecule is in fact too hard an example to carry out in as much detail as one would like, although it illustrates some of the general theory quite nicely. A simpler example for which one can get more detailed information (about relative equilibria and their bifurcations, for example) is the **double spherical pendulum**. Here, instead of the symmetry group being the full (non-abelian) rotation group $SO(3)$, it is the (abelian) group S^1 of rotations about the axis of gravity. The double pendulum will also be used as a thread through the lectures. The results for this example are drawn from Marsden and Scheurle [1992]. To make similar progress with the water molecule, one would have to deal with the already complex issue of finding a reasonable model for the interatomic potential. There is a large literature on this going back to Darling and Dennison [1940] and Sorbie and Murrell [1975]. For some of the recent work that might be important for the present approach, and for more references, see Xiao and Kellman [1989] and Li, Xiao and Kellman [1990].

The special case of the ozone molecule with its three equal masses is also of great interest, not only for environmental reasons, but because this molecule has more symmetry than the water molecule. In fact, what we learn about the water molecule can be used to study the ozone molecule by putting $m = M$. A big change that has very interesting consequences is the fact that the discrete symmetry group is enlarged from "reflections" $\mathbb{Z}_2$ to the "symmetry group of a triangle" D_3. This situation is also of interest in chemistry for things like molecular control by using laser beams to control the potential in which the molecule finds itself. Some believe that, together with ideas from semiclassical quantum mechanics, the study of this system as a classical system provides useful information. We refer to Pierce, Dahleh, and Rabitz [1988], Tannor [1989] and Tannor and Jin [1991] for more information and literature leads.

1.2 Hamiltonian Formulation

The equations of motion for a classical mechanical system with n degrees of freedom may be written as a set of first order equations in Hamiltonian form:

$$\dot{q}^i = \frac{\partial H}{\partial p_i}; \quad \dot{p}^i = -\frac{\partial H}{\partial q_i}; \quad i = 1, \ldots, n. \tag{1.2.1}$$

The configuration coordinates $(q^1, \ldots, q^n)$ and momenta $(p_1, \ldots, p_n)$ together define the system's instantaneous state, which may also be regarded as the coordinates of a point in $\mathbb{R}^{2n}$, the system's **phase space**. We denote such a point by (q, p). The Hamiltonian function $H(q, p)$ defines the system and, in the absence of constraining forces and time dependence, is simply the total energy of the system. The phase space for the water molecule is $\mathbb{R}^{18}$ and the Hamiltonian is the kinetic plus potential energies.

This classical setting can be generalized in two essential ways. First, the phase space may be a higher dimensional surface (*i.e.*, a differentiable manifold) rather than a linear vector space. This generalization allows for the simplest and most natural characterizations of systems consisting of bodies whose motions are spatially constrained and for a deeper under- standing even of an n-particle system like the water molecule. The set of all possible spatial positions of bodies in the system is their *configuration space*. For example, the configuration space for the water molecule is $\mathbb{R}^9$ and for a three dimensional rigid body moving freely in space is $SE(3)$, the six dimensional group of Euclidean (rigid) transformations of three-space, that is, all possible rotations and translations. If translations are ignored and only rotations are considered, then the configuration space is $SO(3)$. As another example, if two rigid bodies are connected at a point by an idealized ball-in-socket joint, then to specify the position of the bodies, we must specify a single translation (since the bodies are coupled) but we need to specify two rotations (since the two bodies are free to rotate in any manner). The configuration space is therefore $SE(3) \times SO(3)$. This is al- ready a fairly complicated object, but remember that one must keep track of both positions and momenta of each component body to formulate the system's dynamics completely. If Q denotes the configuration space (only positions), then the corresponding phase space P (positions and momenta) is the manifold known as the *cotangent bundle* of Q, which is denoted by T^*Q. This is a good way to generalize the phase space of an n-degree of freedom system.

The second important way in which the modern theory of Hamiltonian systems generalizes the classical theory is by relaxing the requirement of using canonical phase space coordinate systems, *i.e.*, coordinate systems in which the equations of motion have the form (1.2.1) above. Rigid body dynamics, celestial mechanics, fluid and plasma dynamics, nonlinear elas- todynamics and robotics provide a rich supply of examples of systems for which canonical coordinates can be unwieldy and awkward. The free mo- tion of a rigid body in space is a basic example of this type. It was treated by Euler in the eighteenth century and yet it remains remarkably rich as an illustrative example. Notice that if our water molecule has *stiff* springs between the atoms, then it behaves nearly like a rigid body. One of our aims is to bring out this behavior.

As mentioned earlier, the rigid body problem in its primitive formula- tion has the six dimensional configuration space $SE(3)$. This means that the phase space, $T^*SE(3)$ is twelve dimensional. Assuming that no external forces act on the body, conservation of linear momentum allows us to solve

for the components of the position and momentum vectors of the center of
mass. Reduction to the center of mass frame, which we will work out in
detail for the classical water molecule, reduces one to the case where the
center of mass is fixed, so only $SO(3)$ remains. Each possible orientation
corresponds to an element of the rotation group $SO(3)$ which we may therefore view as a configuration space for all non-trivial motions of the body.
Euler formulated a description of the body's orientation in space in terms
of three angles between axes that are either fixed in space or are attached
to symmetry planes of the body's motion. The three *Euler angles*, ψ, φ and
θ are generalized coordinates for the problem and form a coordinate chart
for $SO(3)$. However, it is simpler and more convenient to proceed more
intrinsically as follows.

We regard the element $A \in SO(3)$ giving the configuration of the body
as a map of a *reference configuration* $\mathcal{B} \subset \mathbb{R}^3$ to the current configuration $A(\mathcal{B})$ taking a reference or label point $X \in \mathcal{B}$ to a current point
$x = A(X) \in A(\mathcal{B})$. For a rigid body in motion, the matrix A becomes time
dependent and the velocity of a point of the body is $\dot{x} = \dot{A}X = \dot{A}A^{-1}x$.
Since A is an orthogonal matrix, we can write

$$\dot{x} = \dot{A}A^{-1}x = \omega \times x,$$

which defines the *spatial angular velocity*. The corresponding *body
angular velocity* is defined by

$$\Omega = A^{-1}\omega,$$

so that Ω is the angular velocity as seen in a body fixed frame. The kinetic
energy is the usual expression

$$K = \frac{1}{2} \int_{\mathcal{B}} \rho(X) \|\dot{A}X\|^2 d^3X,$$

where ρ is the mass density. Since

$$\|\dot{A}X\| = \|\omega \times x\| = \|A^{-1}(\omega \times x)\| = \|\Omega \times X\|,$$

the kinetic energy is a quadratic function of Ω. Writing

$$K = \frac{1}{2}\Omega^T \mathbb{I}\Omega$$

then defines the *moment of inertia tensor* $\mathbb{I}$, which we can regard as
a positive definite 3×3 matrix, or better, quadratic form. This quadratic

form, can be diagonalized, and this defines the *principal axes and moments of inertia.* In this basis, we write $\mathbb{I} = \text{diag}(I_1, I_2, I_3)$. Every calculus text teaches one how to compute moments of inertia! The *body angular momentum* is defined, analogous to linear momentum $p = mv$, as

$$\Pi = \mathbb{I}\Omega$$

so that in principal axes,

$$\Pi = (\Pi_1, \Pi_2, \Pi_3) = (I_1\Omega_1, I_2\Omega_2, I_3\Omega_3).$$

The equations of motion for the rigid body are the Euler-Lagrange equations for the Lagrangian L equal to the kinetic energy, but regarded as a function on $TSO(3)$ or equivalently, Hamilton's equations with the Hamiltonian equal to the kinetic energy, but regarded as a function on the cotangent bundle of $SO(3)$. In terms of the Euler angles and their conjugate momenta, these are the canonical Hamilton equations, but as such they are a rather complicated set of six ordinary differential equations.

Assuming that no external moments act on the body, the spatial angular momentum vector $\pi = A\Pi$ is conserved in time. As we shall recall in Chapter 2, this follows by general considerations of symmetry. Euler used this to write the three associated momentum equations for the components of the body angular momentum vector obtaining the *Euler equations*:

$$\dot{\Pi}_1 = \frac{I_2 - I_3}{I_2 I_3}\Pi_2\Pi_3$$

$$\dot{\Pi}_2 = \frac{I_3 - I_1}{I_3 I_1}\Pi_3\Pi_1 \qquad\qquad (1.2.2)$$

$$\dot{\Pi}_3 = \frac{I_1 - I_2}{I_1 I_2}\Pi_1\Pi_2$$

Around 1966, it was Arnold who clarified in a satisfactory way the relationships between the various representations (body, space, Euler angles) of the equations and showed how the same ideas apply to fluid mechanics as well; see Arnold [1966].

Viewing (Π_1, Π_2, Π_3) as coordinates in a three dimensional vector space, the Euler equations are evolution equations for a point in this space. An integral (constant of motion) for the system is given by the magnitude of the total angular momentum vector: $\|\Pi\|^2 = \Pi_1^2 + \Pi_1^2 + \Pi_1^2$. This can be verified directly from the Euler equations (1.2.2). Because of this, the evolution in time of any initial point $\Pi(0)$ is constrained to the sphere $\|\Pi\|^2 = \|\Pi(0)\|^2 = \text{constant}$. Thus we may view the Euler equations as describing a two dimensional evolution on an invariant sphere. This sphere is the **reduced phase space** for the rigid body equations. In fact, this defines a two dimensional system as a Hamiltonian dynamical system on the two-sphere S^2. The Hamiltonian structure is not obvious from Euler's equations because the description in terms of the body angular momentum is inherently non-canonical. As we shall see in §1.3 and in more detail in Chapter 4, the theory of Hamiltonian systems may be generalized to include Euler's formulation. The Hamiltonian for the reduced system is

$$H = \frac{1}{2} \left(\frac{\Pi_1^2}{I_1} + \frac{\Pi_2^2}{I_2} + \frac{\Pi_3^2}{I_3} \right) \tag{1.2.3}$$

and we shall show how this function allows us to recover Euler's equations (1.2.2). Since solutions curves of (1.2.2) are confined to the level sets of H (which are in general ellipsoids) as well as to the invariant spheres $\|\Pi\| = $ constant, the intersection of these surfaces are precisely the trajectories of the rigid body, as shown in Figure 1.2.1.

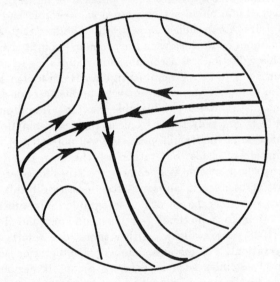

Figure 1.2.1: Phase portrait for the rigid body. The magnitude of the angular momentum vector determines a sphere. The intersection of the sphere with the ellipsoids of constant Hamiltonian gives the trajectories of the rigid body.

On the reduced phase space dynamical fixed points are called **relative equilibria**. These equilibria correspond to periodic orbits in the unreduced phase space, specifically to *steady rotations* about a principal inertial axis. The locations and stability types of the relative equilibria for the rigid body are clear from Figure 1.2.1. The four points located at the intersections of the invariant sphere with the Π_1 and Π_2 axes correspond to pure rotational motions of the body about its major and minor principal axes. These motions are stable, whereas the other two relative equilibria corresponding to rotations about the intermediate principal axis are unstable.

In these lectures we shall see how the stability analysis for a large class of more complicated systems can be simplified through a careful choice of non-canonical coordinates. We managed to visualize the trajectories of the rigid body without doing any calculations, but this occurrence is rare; the rigid body is an especially simple system. Problems like the rotating water molecule will prove to be more challenging. Not only is the rigid body problem integrable (one can write down the solution in terms of integrals), but the problem reduces in some sense to a two dimensional manifold and allows questions about trajectories to be phrased in terms of level sets of integrals. Many Hamiltonian systems are not integrable and trajectories are chaotic and are often studied numerically. The fact that we were able to reduce the number of dimensions in the problem (from twelve to two) and the fact that this reduction was accomplished by appealing to non-canonical coordinates turns out to be a general feature for Hamiltonian systems with symmetry. The reduction procedure may be applied to non-integrable or chaotic systems, just as well as to integrable ones. In a Hamiltonian context, non-integrability is generally taken to mean that any analytic constant of motion is a function of the Hamiltonian. We will not attempt to formulate a general definition of chaos, but rather use the term in a loose way to refer to systems whose motion is so complicated that long-term prediction of dynamics is impossible. It can sometimes be very difficult to establish whether a given system is chaotic or non-integrable. Sometimes theoretical tools such as "Melnikov's method" (see Guckenheimer and Holmes [1983] and Wiggins [1988]) are available. Other times, one resorts to numerics or direct observation. For instance, numerical integration suggests that irregular natural satellites such as Saturn's moon, Hyperion, tumble in their orbits in a highly irregular manner (see Wisdom, Peale and Mignard [1984]). The equations of motion for an irregular body in the presence of a non-uniform gravitational field are similar to the Euler equations except that there is a configuration-dependent gravitational moment term in the equations that may render the system non-integrable.

The evidence that Hyperion tumbles chaotically in space leads to difficulties in numerically modelling this system. The manifold $SO(3)$ cannot be covered by a single three dimensional coordinate chart such as the Euler angle chart. We shall prove this in §1.6. Hence an integration algorithm using canonical variables must employ more than one coordinate system, alternating between coordinates on the basis of the body's current configuration. For a body that tumbles in a complicated fashion, the body's configuration might switch from one chart of $SO(3)$ to another in a short time interval, and the computational cost for such a procedure could be prohibitive for long time integrations. This situation is worse still for bodies with internal degrees of freedom like our water molecule, robots, and large-scale space structures. Such examples point out the need to go beyond canonical formulations.

1.3 Geometry, Symmetry, and Reduction

We have emphasized the distinction between canonical and non-canonical coordinates by contrasting Hamilton's (canonical) equations with Euler's equations. We may view this distinction from a different perspective by introducing Poisson bracket notation. Given two smooth (C^∞) real-valued functions F and K defined on the phase space of a Hamiltonian system, define the *canonical Poisson bracket* of F and K by

$$\{F, K\} = \sum_{i=1}^{n} \left(\frac{\partial F}{\partial q^i} \frac{\partial K}{\partial p_i} - \frac{\partial K}{\partial q^i} \frac{\partial F}{\partial p_i} \right) \tag{1.3.1}$$

where (q^i, p_i) are conjugate pairs of canonical coordinates. If H is the Hamiltonian function for the system, then the formula for the Poisson bracket is the directional derivative of F along the flow of Hamilton's equations; that is,

$$\dot{F} = \{F, H\}. \tag{1.3.2}$$

In particular, Hamilton's equations are recovered if we let F be each of the canonical coordinates in turn:

$$\dot{q}^i = \{q^i, H\} = \frac{\partial H}{\partial p_i}, \quad \dot{p}_i = \{p_i, H\} = -\frac{\partial H}{\partial q^i}.$$

Once H is specified, the chain rule shows that the statement "$\dot{F} = \{F, H\}$ for all smooth functions F" is equivalent to Hamilton's equations. In fact, it tells how any function F evolves along the flow.

This representation of the canonical equations of motion suggests a generalization of the bracket notation to cover non-canonical formulations. As an example, consider Euler's equations (1.2.2). Define the following non-canonical *rigid body bracket* of two smooth functions F and K on the angular momentum space:

$$\{F, K\} = -\Pi \cdot (\nabla F \times \nabla K), \tag{1.3.3}$$

where $\{F, K\}$ and the gradients are evaluated at $\Pi = (\Pi_1, \Pi_2, \Pi_3)$. The notation in (1.3.3) is that standard scalar triple product operation in $\mathbb{R}^3$. If H is the rigid body Hamiltonian (see (1.2.3)) and F is, in turn, allowed to be each of the three coordinate functions Π_i, then the formula $\dot{F} = \{F, H\}$ yields the three Euler equations.

The non-canonical bracket corresponding to the reduced free rigid body problem is an example of what is known as a *Lie-Poisson* bracket. Other bracket operations have been developed to handle a wide variety of Hamiltonian problems in non-canonical form, including some problems outside of the framework of traditional Newtonian mechanics (see for instance, Arnold [1966] or Marsden et al., [1983]). In Hamiltonian dynamics, it is essential to distinguish features of the dynamics that depend on the Hamiltonian function from those that depend only on properties of the phase space. The generalized bracket operation is a geometric invariant in the sense that it depends only on the structure of the phase space. The phase spaces arising in mechanics often have an additional geometric structure closely related to the Poisson bracket. Specifically, they may be equipped with a special differential two-form called the *symplectic form*. The symplectic form defines the geometry of a symplectic manifold much as the metric tensor defines the geometry of a Riemannian manifold. Bracket operations can be defined entirely in terms of the symplectic form without reference to a particular coordinate system.

The classical concept of a canonical transformation can also be given a more geometric definition within this framework. A canonical transformation is classically defined as a transformation of phase space that takes one canonical coordinate system to another. The modern analogue of this concept is a *symplectic map*, a smooth map of a symplectic manifold to itself that preserves the symplectic form or, equivalently, the Poisson bracket operation.

The geometry of symplectic manifolds is an essential ingredient in the formulation of the reduction procedure for Hamiltonian systems with symmetry. We now outline some important ingredients of this procedure and will go into this in more detail in Chapters 2 and 3. In Euler's problem of the free rotation of a rigid body in space (assuming that we have already exploited conservation of linear momentum), the six dimensional phase space is $T^*SO(3)$ — the cotangent bundle of the three dimensional rotation group. This phase space $T^*SO(3)$ is often parametrized by three Euler angles and their conjugate momenta. The reduction from six to two dimensions is a consequence of two essential features of the problem:

1. Rotational invariance of the Hamiltonian, and

2. The existence of a corresponding conserved quantity, μ, the *spatial*

angular momentum.

These two conditions are generalized to arbitrary mechanical systems with symmetry in the general reduction theory of Meyer [1973] and Marsden and Weinstein [1974], which was inspired by the seminal works of Arnold [1966] and Smale [1970]. In this theory, one begins with a given phase space that we denote by P. We assume there is a group G of symmetry transformations of P that transform P to itself by canonical transformation. Generalizing 2, we use the symmetry group to generate a vector-valued conserved quantity denoted $\mathbf{J}$ and called the *momentum map*.

Analogous to the set where the total angular momentum has a given value, we consider the set of all phase space points where $\mathbf{J}$ has a given value μ; *i.e.*, the μ-level set for $\mathbf{J}$. The analogue of the two dimensional body angular momentum sphere in Figure 1.2.1, is the reduced phase space, denoted P_μ that is constructed as follows:

> P_μ *is the μ-level set for* $\mathbf{J}$ *on which any two points that can be transformed one to the other by a group transformation are identified.*

The *reduction theorem* states that

> P_μ *inherits the symplectic (or Poisson bracket) structure from that of P, so it can be used as a new phase space. Also, dynamical trajectories of the Hamiltonian H on P determine corresponding trajectories on the reduced space.*

This new dynamical system is, naturally, called the *reduced system*. The trajectories on the sphere in Figure 1.2.1 are the reduced trajectories for the rigid body problem.

We saw that steady rotations of the rigid body correspond to fixed points on the reduced manifold, namely the body angular momentum sphere. In general, fixed points of the reduced dynamics on P_μ are called *relative equilibria*, following terminology introduced by Poincaré around 1880. The reduction process can be applied to the system that models the motion of the moon Hyperion, to spinning tops, to fluid and plasma systems, and to systems of coupled rigid bodies. For example, if our water molecule is undergoing steady rotation, with the internal parts not moving relative to each other, this will be a relative equilibrium of the system. An oblate Earth in steady rotation is a relative equilibrium for a fluid-elastic body. In general, the bigger the symmetry group, the richer the supply of relative equilibria.

1.4   Stability

There is a standard procedure for finding the stability of equilibria of an ordinary differential equation

$$\dot{x} = f(x) \qquad\qquad (1.4.1)$$

where $x = (x^1, \ldots, x^n)$ and f is smooth. Equilibria are points x_e such that $f(x_e) = 0$; *i.e.*, points that are fixed in time under the dynamics. The goal is to determine the stability of the fixed point x_e. By stability we mean that *any solution to $\dot{x} = f(x)$ that starts near x_e remains close to x_e for all future time.* A traditional method of ascertaining the stability of x_e is to examine the first variation equation

$$\dot{\xi} = \mathbf{D}_x f(x_e)\xi \qquad\qquad (1.4.2)$$

where $\mathbf{D}_x f(x_e)$ is the Jacobian of f at x_e, defined to be the matrix of partial derivatives

$$\mathbf{D}_x f(x_e) = \left[\frac{\partial f^i}{\partial x^j}\right]_{x=x_e}. \qquad\qquad (1.4.3)$$

Liapunov's theorem. *If all the eigenvalues of $\mathbf{D}_x f(x_e)$ lie in the strict left half plane, then the fixed point is stable. If any of the eigenvalues lie in the right half plane, then the fixed point is unstable.*

For Hamiltonian systems, the eigenvalues come in quartets that are symmetric about the origin, and so they cannot all lie in the strict left half plane. (See, for example, Abraham and Marsden [1978] for the proof of this assertion). Thus, *the above form of Liapunov's theorem is not appropriate to deduce whether or not a fixed point of a Hamiltonian system is stable.*

When the Hamiltonian is in canonical form, one can use a stability test for fixed points due to Lagrange and Dirichlet. This method starts with the observation that for a fixed point (q_e, p_e) of such a system,

$$\frac{\partial H}{\partial q}(q_e, p_e) = \frac{\partial H}{\partial p}(q_e, p_e) = 0.$$

Hence the *fixed point occurs at a critical point of the Hamiltonian.*

Lagrange-Dirichlet Criterion. *If the $2n \times 2n$ matrix $\delta^2 H$ of second partial derivatives, (the second variation) is either positive or negative definite at (q_e, p_e) then it is a stable fixed point.*

The proof is very simple. Consider the positive definite case. Since H has a nondegenerate minimum at $z_e = (q_e, p_e)$, Taylor's theorem with remainder shows that its level sets near z_e are bounded inside and outside by spheres of arbitrarily small radius. Since energy is conserved, solutions stay on level surfaces of H, so a solution starting near the minimum has to stay near the minimum.

For a Hamiltonian of the form kinetic plus potential V, critical points occur when $p_e = 0$ and q_e is a critical point of the potential of V. The Lagrange-Dirichlet Criterion then reduces to asking for a non-degenerate minimum of V.

In fact, this criterion was used in one of the classical problems of the 19th century: the problem of rotating gravitating fluid masses. This problem was studied by Newton, MacLaurin, Jacobi, Riemann, Poincaré and others. The motivation for its study was in the conjectured birth of two planets by the splitting of a large mass of solidifying rotating fluid. Riemann [1860], Routh [1877] and Poincaré [1885, 1892, 1901] were major contributors to the study of this type of phenomenon and used the potential energy and angular momentum to deduce the stability and bifurcation.

The Lagrange-Dirichlet method was adapted by Arnold [1966, 1969] into what has become known as the *energy-Casimir method*. Arnold analyzed the stability of stationary flows of perfect fluids and arrived at an explicit stability criterion when the configuration space Q for the Hamiltonian of this system is the symmetry group G of the mechanical system.

A *Casimir function* C is one that Poisson commutes with any function f defined on the phase space of the Hamiltonian system, *i.e.*,

$$\{C, F\} = 0. \tag{1.4.4}$$

Large classes of Casimirs can occur when the reduction procedure is performed, resulting in systems with non-canonical Poisson brackets. For example, in the case of the rigid body discussed previously, if Φ is a function of one variable and μ is the angular momentum vector in the inertial coordinate system, then

$$C(\mu) = \Phi(\|\mu\|^2) \tag{1.4.5}$$

is readily checked to be a Casimir for the rigid body bracket (1.3.3).

Energy-Casimir method. *Choose C such that $H + C$ has a critical point at an equilibrium z_e and compute the second variation $\delta^2(H + C)(z_e)$. If this matrix is positive or negative definite, then the equilibrium z_e is stable.*

When the phase space is obtained by reduction, the equilibrium z_e is called a *relative equilibrium* of the original Hamiltonian system.

The energy-Casimir method has been applied to a variety of problems including problems in fluids and plasmas (Holm et al. [1985]) and rigid bodies with flexible attachments (Krishnaprasad and Marsden [1987]). If applicable, the energy-Casimir method may permit an explicit determination of the stability of the relative equilibria. It is important to remember, however, that these techniques give stability information only. As such one cannot use them to infer instability without further investigation.

The energy-Casimir method is restricted to certain types of systems, since its implementation relies on an abundant supply of Casimir functions. In some important examples, such as the dynamics of geometrically exact flexible rods, Casimirs have not been found and may not even exist. A method developed to overcome this difficulty is known as the *energy momentum method*, which is closely linked to the method of reduction. It uses conserved quantities, namely the energy and momentum map, that are readily available, rather than Casimirs.

The energy momentum method (Marsden, Simo, Lewis and Posbergh [1989], Simo, Posbergh and Marsden [1990, 1991], Simo, Lewis and Marsden [1991], and Lewis and Simo [1990]) involves the *augmented Hamiltonian* defined by

$$H_\xi(q,p) = H(q,p) - \xi \cdot \mathbf{J}(q,p) \tag{1.4.6}$$

where $\mathbf{J}$ is the momentum map described in the previous section and ξ may be thought of as a Lagrange multiplier. For the water molecule, $\mathbf{J}$ is the angular momentum and ξ is the angular velocity of the relative equilibrium. One sets the first variation of H_ξ equal to zero to obtain the relative equilibria. To ascertain stability, the second variation $\delta^2 H_\xi$ is calculated. One is then interested in determining the definiteness of the second variation.

Definiteness in this context has to be properly interpreted to take into account the conservation of the momentum map $\mathbf{J}$ and the fact that $\mathbf{D}^2 H_\xi$ may have zero eigenvalues due to its invariance under a subgroup of the symmetry group. The variations of p and q must satisfy the linearized angular momentum constraint $(\delta q, \delta p) \in \ker[\mathbf{DJ}(q_e, p_e)]$, and must not lie in symmetry directions; only these variations are used to calculate the second variation of the augmented Hamiltonian H_ξ. These define the space of *admissible variations* $\mathcal{V}$. The energy momentum method has been applied to the stability of relative equilibria of among others, geometrically exact rods and coupled rigid bodies (Patrick [1989, 1990] and Simo, Posbergh and Marsden [1990, 1991]).

A cornerstone in the development of the energy-momentum method was laid by Routh [1877] and Smale [1970] who studied the stability of relative equilibria of simple mechanical systems. Simple mechanical systems are those whose Hamiltonian may be written as the sum of the potential and kinetic energies. Part of Smale's work may be viewed as saying that there is a naturally occuring connection called the **mechanical connection** on the reduction bundle that plays an important role. A connection can be thought of as a generalization of the electromagnetic vector potential.

The **amended potential** V_μ is the potential energy of the system plus a generalization of the potential energy of the centrifugal forces in stationary rotation:

$$V_\mu(q) = V(q) + \frac{1}{2}\mu \cdot \mathbb{I}^{-1}(q)\mu \qquad (1.4.7)$$

where $\mathbb{I}$ is the **locked inertia tensor**, a generalization of the inertia tensor of the rigid structure obtained by locking all joints in the configuration q. We will define it precisely in Chapter 3 and compute it for several examples. Smale showed that relative equilibria are critical points of the amended potential V_μ, a result we prove in Chapter 4. The corresponding momentum p need not be zero since the system is typically in motion.

The second variation $\delta^2 V_\mu$ of V_μ directly yields the stability of the relative equilibria. However, an interesting phenomenon occurs if the space $\mathcal{V}$ of admissible variations is split into two specially chosen subspaces $\mathcal{V}_{\text{RIG}}$ and $\mathcal{V}_{\text{INT}}$. In this case the second variation *block diagonalizes*:

$$\delta^2 V_\mu \mid \mathcal{V} \times \mathcal{V} = \begin{bmatrix} D^2 V_\mu \mid \mathcal{V}_{\text{RIG}} \times \mathcal{V}_{\text{RIG}} & 0 \\ & \\ 0 & D^2 V_\mu \mid \mathcal{V}_{\text{INT}} \times \mathcal{V}_{\text{INT}} \end{bmatrix} \qquad (1.4.8)$$

The space $\mathcal{V}_{\text{RIG}}$ (*rigid variations*) is generated by the symmetry group, and $\mathcal{V}_{\text{INT}}$ are the **internal** or **shape variations**. In addition, the whole matrix $\delta^2 H_\xi$ block diagonalizes in a very efficient manner as we will see in Chapter 5. This often allows the stability conditions associated with $\delta^2 V_\mu \mid \mathcal{V} \times \mathcal{V}$ to be recast in terms of a standard eigenvalue problem for the second variation of the amended potential.

This splitting *i.e.*, block diagonalization, has more miracles associated with it. In fact,

> *the second variation $\delta^2 H_\xi$ and the symplectic structure (and therefore the equations of motion) can be explicitly brought into normal form simultaneously.*

This result has several interesting implications. In the case of pseudo-rigid bodies (Lewis and Simo [1990]), it reduces the stability problem from an unwieldy 14×14 matrix to a relatively simple 3×3 subblock on the diagonal. The block diagonalization procedure enabled Lewis and Simo to solve their problem analytically, whereas without it, a substantial numerical computation would have been necessary.

As we shall see in Chapter 8, the presense of discrete symmetries (as for the water molecule and the pseudo-rigid bodies) gives further, or refined, subblocking properties in the second variation of $\delta^2 H_\xi$ and $\delta^2 V_\mu$ and the symplectic form.

In general, this diagonalization explicitly separates the rotational and internal modes, a result which is important not only in rotating and elastic fluid systems, but also in molecular dynamics and robotics. Similar simplifications are expected in the analysis of other problems to be tackled using the energy momentum method.

1.5 Geometric Phases

The application of the methods described above is still in its infancy, but the previous example indicates the power of reduction and suggests that the energy-momentum method will be applied to dynamic problems in many fields, including chemistry, quantum and classical physics, and engineering. Apart from the computational simplification afforded by reduction, reduction also permits us to put into a mechanical context a concept known as the *geometric phase*, or *holonomy*.

An example in which holonomy occurs is the Foucault pendulum. During a single rotation of the earth, the plane of the pendulum's oscillations is shifted by an angle that depends on the latitude of the pendulum's location. Specifically if a pendulum located at co-latitude (*i.e.*, the azimuthal angle.) α is swinging in a plane, then after twenty-four hours, the plane of its oscillations will have shifted by an angle $2\pi \cos \alpha$. This holonomy is (in a non-obvious way) a result of parallel translation: if an orthonormal coordinate frame undergoes parallel transport along a line of co-latitude α, then after one revolution the frame will have rotated by an amount equal to the phase shift of the Foucault pendulum (see Figure 1.5.1).

Geometrically, the holonomy of the Foucault pendulum is equal to the solid angle swept out by the pendulum's axis during one rotation of the earth. Thus a pendulum at the north pole of the earth will experience a holonomy of 2π. If you imagine parallel transporting a vector around a small loop near the north pole, it is clear that one gets an answer close to 2π, which agrees with what the pendulum experiences. On the other hand, a pendulum on the earth's equator experiences no holonomy.

A less familiar example of holonomy was presented by Hannay [1985] and discussed further by Berry [1985]. Consider a frictionless, non-circular, planar hoop of wire on which is placed a small bead. The bead is set in motion and allowed to slide along the wire at a constant speed (see Figure 1.5.2). (We will need the notation in this figure only later in Chapter

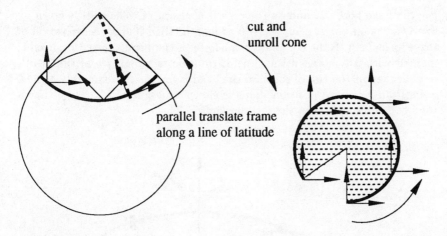

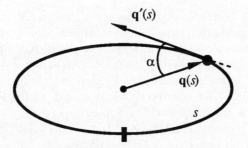

Figure 1.5.1: The parallel transport of a coordinate frame along a curved surface.

6.) Clearly the bead will return to its initial position after, say, τ seconds, and will continue to return every τ seconds after that. Allow the bead to make many revolutions along the circuit, but for a fixed amount of total time, say T.

Figure 1.5.2: A bead sliding on a planar, non-circular hoop of area A and length L. The bead slides around the hoop at constant speed with period τ and is allowed to revolve for time T.

Suppose that the wire hoop is *slowly* rotated in its plane by 360 degrees while the bead is in motion for exactly the same total length of time T. At the end of the rotation, the bead is not in the location where we might expect it, but instead will be found at a shifted position that is determined by the shape of the hoop. In fact, the shift in position depends only on the length of the hoop, L, and on the area it encloses, A. The shift is given by $8\pi^2 A/L^2$ as an angle, or by $4\pi A/L$ as length. (See §6.6 for a derivation of these formulas.) To be completely concrete, if the bead's initial position is marked with a tick and if the time of rotation is a multiple of the bead's period, then at the end of rotation the bead is found approximately $4\pi A/L$ units from its initial position. This is shown in Figure 1.5.3. Note that if the hoop is circular then the angular shift is 2π.

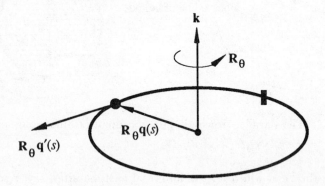

Figure 1.5.3: The hoop is slowly rotated in the plane through 360 degrees. After one rotation, the bead is located $4\pi A/L$ units behind where it would have been, had the rotation not occurred.

Let us indicate how holonomy is linked to the reduction process by returning to our rigid body example. The rotational motion of a rigid body can be described as a geodesic (with respect to the inertia tensor regarded as a metric) on the manifold $SO(3)$. As mentioned earlier, for each angular momentum vector μ, the reduced space P_μ can be identified with the two-sphere of radius $\|\mu\|$. This construction corresponds to the Hopf fibration which describes the three-sphere S^3 as a nontrivial circle bundle over S^2. In our example, S^3 (or rather $S^3/\mathbb{Z}_2 \cong \mathbf{J}^{-1}(\mu)$) is the subset of phase space which is mapped to μ under the reduction process.

Suppose we are given a trajectory $\Pi(t)$ on P_μ that has period T and energy E. Following Montgomery [1991] and Marsden, Montgomery and Ratiu [1990] we shall show in §6.4 that after time T the rigid body has rotated in physical 3-space about the axis μ by an angle (modulo 2π)

$$\Delta\theta = -\Lambda + \frac{2ET}{\|\mu\|}. \tag{1.5.1}$$

Here Λ is the solid angle subtended by the closed curve $\Pi(t)$ on the sphere S^2 and is oriented according to the right hand rule. The *approximate* phase formula $\Delta\theta \cong 8\pi^2 A/L^2$ for the ball in the hoop is derived by the classical techniques of averaging and the variation of constants formula. However, formula (1.5.1) is *exact*. (In Whittaker [1959], (1.5.1) is expressed as a complicated quotient of theta functions!)

An interesting feature of (1.5.1) is the manner in which $\Delta\theta$ is split into two parts. The term Λ is purely geometric and so is called the *geometric phase*. It does not depend on the energy of the system or the period of motion, but rather on the fraction of the surface area of the sphere P_μ that is enclosed by the trajectory $\Pi(t)$. The second term in (1.5.1) is known as the *dynamic phase* and depends explicitly on the system's energy and the period of the reduced trajectory.

Geometrically we can picture the rigid body as tracing out a path in its phase space. More precisely, conservation of angular momentum implies that the path lies in the submanifold consisting of all points that are mapped onto μ by the reduction process. As Figure 1.2.1 shows, almost every trajectory on the reduced space is periodic, but this does not imply that the original path was periodic, as is shown in Figure 1.5.4. The difference between the true trajectory and a periodic trajectory is given by the holonomy plus the dynamic phase.

It is possible to observe the holonomy of a rigid body with a simple experiment. Put a rubber band around a book so that the cover will not open. (A "tall", thin book works best.) With the front cover pointing up, gently toss the book in the air so that it rotates about its middle axis (see Figure 1.5.5). Catch the book after a single rotation and you will find that it has also rotated by 180 degrees about its long axis — that is, the front cover is now facing the floor!

This particular phenomena is not *literally* covered by Montgomery's formula since we are working close to the homoclinic orbit and in this limit $\Delta\theta \to +\infty$ due to the limiting steady rotations. Thus, "catching" the book plays a role. For an analysis from another point of view, see Ashbaugh, Chicone and Cushman [1991].

There are other everyday occurrences that demonstrate holonomy. For example, a falling cat usually manages to land upright if released upside down from complete rest, that is, with total angular momentum zero. This ability has motivated several investigations in physiology as well as dynamics and more recently has been analyzed by Montgomery [1990] with an emphasis on how the cat (or, more generally, a deformable body) can efficiently readjust its orientation by changing its shape. By "efficiently", we mean that the reorientation minimizes some function — for example the total energy expended. In other words, one has a problem in *optimal control*. Montgomery's results characterize the deformations that allow a cat to reorient itself without violating conservation of angular momentum.

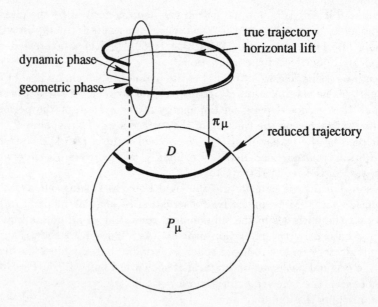

Figure 1.5.4: Holonomy for the rigid body. As the body completes one period in the reduced phase space P_μ, the body's true configuration does not return to its original value. The phase difference is equal to the influence of a dynamic phase which takes into account the body's energy, and a geometric phase which depends only on the area of P_μ enclosed by the reduced trajectory.

In his analysis, Montgomery casts the falling cat problem into the language of principal bundles. Let the shape of a cat refer to the location of the cat's body parts relative to each other, but without regard to the cat's orientation in space. Let the configuration of a cat refer both to the cat's shape and to its orientation with respect to some fixed reference frame. More precisely, if Q is the configuration space and G is the group of rigid motions, then Q/G is the *shape space*.

If the cat is completely rigid then it will always have the same shape, but we can give it a different configuration by rotating it through, say 180 degrees about some axis. If we require that the cat has the same shape at the end of its fall as it had at the beginning, then the cat problem may be formulated as follows: Given an initial configuration, what is the most efficient way for a cat to achieve a desired final configuration if the final shape is required to be the same as the initial shape? If we think of the cat as tracing out some path in configuration space during its fall, the

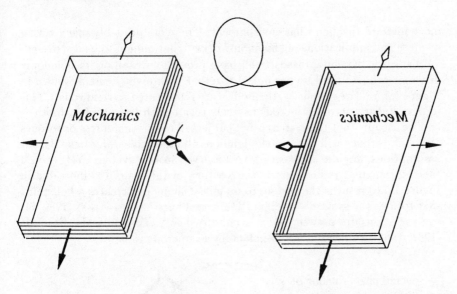

Figure 1.5.5: A book tossed in the air about an axis that is close to the middle (unstable) axis experiences a holonomy of 180 degrees about its long axis when caught after one revolution.

projection of this path onto the shape space results in a trajectory in the shape space, and the requirement that the cat's initial and final shapes are the same means that the trajectory is a closed loop. Furthermore, if we want to know the most efficient configuration path that satisfies the initial and final conditions, then we want to find the shortest path with respect to a metric induced by the function we wish to minimize. It turns out that the solution of a falling cat problem is closely related to Wong's equations that describe the motion of a colored particle in a Yang-Mills field (Montgomery [1990], Shapere and Wilczek [1989]). We will come back to these points in Chapter 7.

The examples above indicate that holonomic occurrences are not rare. In fact, Shapere and Wilczek showed that aquatic microorganisms use holonomy as a form of propulsion. Because these organisms are so small, the environment in which they live is extremely viscous to them. The apparent viscosity is so great, in fact, that they are unable to swim by conventional stroking motions, just as a person trapped in a tar pit would be unable to swim to safety. These microorganisms surmount their locomotion difficulties, however, by moving their "tails" or changing their shapes in a topologically nontrivial way that induces a holonomy and allows them to move forward through their environment. There are probably many consequences and applications of "holonomy drive" that remain to be discovered.

Yang and Krishnaprasad [1990] have provide an example of holonomy drive for coupled rigid bodies linked together with pivot joints as shown in Figure 1.5.5. (For simplicity, the bodies are represented as rigid rods.) This form of linkage permits the rods to freely rotate with respect of each other, and we assume that the system is not subjected to external forces or torques although torques will exist in the joints as the assemblage rotates. By our assumptions, angular momentum is conserved in this system. Yet, even if the total angular momentum is zero, a turn of the crank (as indicated in Figure 1.5.6) returns the system to its initial shape but creates a holonomy that rotates the system's configuration. See Thurston and Weeks [1986] for some relationships between linkages and the theory of 3-manifolds. Brockett [1987, 1989] studies the use of holonomy in micromotors.

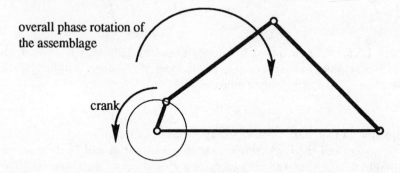

Figure 1.5.6: Rigid rods linked by pivot joints. As the "crank" traces out the path shown, the assemblage experiences a holonomy resulting in a clockwise shift in its configuration.

Holonomy also comes up in the field of magnetic resonance imaging (MRI) and spectroscopy. Berry's work shows that if a quantum system experiences a slow (adiabatic) cyclic change, then there will be a shift in the phase of the system's wave function. This is a quantum analogue to the bead on a hoop problem discussed above. This work has been verified by several independent experiments; the implications of this result to MRI and spectroscopy are still being investigated. For a review of the applications of the geometric phase to the fields of spectroscopy, field theory, and solid-state physics, see Zwanziger, Koenig and Pines [1990] and the bibliography therein.

Another possible application of holonomy drive is to the somersaulting robot. Due to the finite precision response of motors and actuators, a slight error in the robot's initial angular momentum can result in an unsatisfactory landing as the robot attempts a flip. Yet, in spite of the challenges, Hodgings and Raibert [1990] report that the robot can execute 90 percent of the flips successfully. Montgomery, Raibert and Li [1990] are asking whether a robot can use holonomy to improve this rate of success. To do this, they reformulate the falling cat problem as a problem in feedback control: the cat must use information gained by its senses in order to determine how to twist and turn its body so that it successfully lands on its feet.

It is possible that the same technique used by cats can be implemented in a robot that also wants to complete a flip in mid-air. Imagine a robot installed with sensors so that as it begins its somersault it measures its momenta (linear and angular) and quickly calculates its final landing position. If the calculated final configuration is different from the intended final configuration, then the robot waves mechanical arms and legs while entirely in the air to create a holonomy that equals the difference between the two configurations.

If "holonomy drive" can be used to control a mechanical structure, then there may be implications for future satellites like a space telescope. Suppose the telescope initially has zero angular momentum (with respect to its orbital frame), and suppose it needs to be turned 180 degrees. One way to do this is to fire a small jet that would give it angular momentum, then, when the turn is nearly complete, to fire a second jet that acts as a brake to exactly cancel the aquired angular momentum. As in the somersaulting robot, however, errors are bound to occur, and the process of returning the telescope to (approximately) zero angular momentum may be a long process. It would seem to be more desirable to turn it while constantly preserving zero angular momentum. The falling cat performs this very trick. A telescope can mimic this with internal momentum wheels or with flexible joints. This brings us to the area of control theory and its relation to the present ideas. Bloch, Krishnaprasad, Marsden and Sánchez de Alvarez [1991] represents one step in this direction. We shall also discuss their result in Chapter 7.

1.6 The Rotation Group and the Poincaré Sphere

The rotation group $SO(3)$, consisting of all 3×3 orthogonal matrices with determinant one, plays an important role for problems of interest in this book, and so one should try to understand it a little more deeply. As a first try, one can contemplate Euler's Theorem, which states that every rotation in $\mathbb{R}^3$ is a rotation through *some angle* about *some axis*. While true, this can be misleading if not used with care. For example, it suggests that we can identify the set of rotations with the set consisting of all unit vectors in $\mathbb{R}^3$ (the axes) and numbers between 0 and 2π (the angles); that is, with the set $S^2 \times S^1$. However, *this is false* for reasons that involve some basic topology. Thus, a better approach is needed.

One method for gaining deeper insight is to realize $SO(3)$ as $SU(2)/\mathbb{Z}_2$ (where $SU(2)$ is the group of 2×2 complex unitary matrices of determinant 1), using quaternions and Pauli spin matrices; see Abraham and Marsden [1978], p. 273-4, for the precise statement. This approach also shows that the group $SU(2)$ is diffeomorphic to the set of unit quaternions, the three sphere S^4 in $\mathbb{R}^4$.

The **Hopf fibration** is the map of S^3 to S^2 defined, using the above approach to the rotation group, as follows. First map a point $w \in S^3$ to its equivalence class $A = [w] \in SO(3)$ and then map this point to $A\mathbf{k}$, where $\mathbf{k}$ is the standard unit vector in $\mathbb{R}^3$ (or any other fixed unit vector in $\mathbb{R}^3$).

Closely related to the above description of the rotation group, but in fact a little more straightforward, is Poincaré's representation of $SO(3)$ as the unit circle bundle $T_1 S^2$ of the two sphere S^2. This comes about as follows. Elements of $SO(3)$ can be identified with oriented orthonormal frames in $\mathbb{R}^3$; *i.e.*, with triples of orthonormal vectors $(\mathbf{n}, \mathbf{m}, \mathbf{n} \times \mathbf{m})$. Such triples are in one to one correspondence with the points in $T_1 S^2$ by $(\mathbf{n}, \mathbf{m}, \mathbf{n} \times \mathbf{m}) \leftrightarrow (\mathbf{n}, \mathbf{m})$ where $\mathbf{n}$ is the base point in S^2 and $\mathbf{m}$ is regarded as a vector tangent to S^2 at the point $\mathbf{n}$ (see Figure 1.6.1).

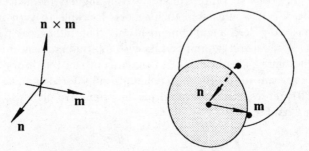

Figure 1.6.1: The rotation group is diffeomorphic to the unit tangent bundle to the two sphere.

Poincaré's representation shows that $SO(3)$ *cannot* be written as $S^2 \times S^1$ (*i.e.*, one *cannot* globally and in a unique and singularity free way, write rotations using an *axis* and an *angle*, despite Euler's theorem). This is because of the (topological) fact that every vector field on S^2 must vanish somewhere. Thus, we see that this is closely related to the nontriviality of the Hopf fibration.

Not only does Poincaré's representation show that $SO(3)$ is topologically non-trivial, but this representation is *useful* in mechanics. In fact, Poincare's representation of $SO(3)$ as the unit circle bundle $T_1 S^2$ is the stepping stone in the formulation of optimal singularity free parameterizations for geometrically exact plates and shells. These ideas have been exploited within a numerical analysis context in Simo, Fox and Rifai [1990] for statics and [1991] for dynamics. Also, this representation is helpful in studying the problem of reorienting a rigid body using internal momentum wheels. We shall treat some aspects of these topics in the course of the lectures.

Chapter 2

A Crash Course in Geometric Mechanics

We now set out the notation and terminology used in subsequent lectures. The reader is referred to one of the standard books, such as Abraham and Marsden [1978], Arnold [1989], Guillemin and Sternberg [1984] and Marsden and Ratiu [1992] for proofs omitted here.[1]

2.1 Symplectic and Poisson Manifolds

Definition 2.1 *Let P be a manifold and let $\mathcal{F}(P)$ denote the set of smooth real-valued functions on P. Consider a given bracket operation denoted*

$$\{\,,\} : \mathcal{F}(P) \times \mathcal{F}(P) \to \mathcal{F}(P).$$

*The pair $(P, \{\,,\})$ is called a **Poisson manifold** if $\{\,,\}$ satisfies*

(PB1)	*bilinearity*	$\{f, g\}$ *is bilinear in f and g.*
(PB2)	*anticommutativity*	$\{f, g\} = -\{g, f\}$.
(PB3)	*Jacobi's identity*	$\{\{f, g\}h\} + \{\{h, f\}g\} + \{\{g, h\}f\} = 0$.
(PB4)	*Leibniz' rule*	$\{fg, h\} = f\{g, h\} + g\{f, h\}$.

Conditions (PB1) – (PB3) make $(\mathcal{F}(P), \{\,,\})$ into a Lie algebra. If $(P, \{\,,\})$ is a Poisson manifold, then because of (PB1) and (PB4), there is a tensor B on P, assigning to each $z \in P$ a linear map $B(z) : T_z^* P \to T_z P$ such that

$$\{f, g\}(z) = \langle B(z) \cdot \mathbf{d}f(z), \mathbf{d}g(z) \rangle. \tag{2.1.1}$$

---

[1] We thank Hans Peter Kruse for providing a helpful draft of the notes for this lecture.

Here, $\langle \, , \rangle$ denotes the natural pairing between vectors and covectors. Because of (PB2), $B(z)$ is antisymmetric. Letting $z^I, I = 1, \ldots, M$ denote coordinates on P, (2.1.1) becomes

$$\{f, g\} = B^{IJ} \frac{\partial f}{\partial z^I} \frac{\partial g}{\partial z^J}. \tag{2.1.2}$$

Antisymmetry means $B^{IJ} = -B^{JI}$ and Jacobi's identity reads

$$B^{LI} \frac{\partial B^{JK}}{\partial z^L} + B^{LJ} \frac{\partial B^{KI}}{\partial z^L} + B^{LK} \frac{\partial B^{IJ}}{\partial z^L} = 0. \tag{2.1.3}$$

Definition 2.2 *Let* $(P_1, \{\,,\}_1)$ *and* $(P_2, \{\,,\}_2)$ *be Poisson manifolds. A mapping* $\varphi : P_1 \to P_2$ *is called **Poisson** if for all* $f, h \in \mathcal{F}(P_2)$, *we have*

$$\{f, h\}_2 \circ \varphi = \{f \circ \varphi, h \circ \varphi\}_1. \tag{2.1.4}$$

Definition 2.3 *Let* P *be a manifold and* Ω *a 2-form on* P. *The pair* (P, Ω) *is called a **symplectic manifold** if* Ω *satisfies*

(S1) $\mathbf{d}\Omega = 0$ *(i.e., Ω is closed) and*
(S2) Ω *is nondegenerate.*

Definition 2.4 *Let* (P, Ω) *be a symplectic manifold and let* $f \in \mathcal{F}(P)$. *Let* X_f *be the unique vector field on* P *satisfying*

$$\Omega_z(X_f(z), v) = \mathbf{d}f(z) \cdot v \quad \text{for all} \quad v \in T_z P. \tag{2.1.5}$$

We call X_f *the **Hamiltonian vector field** of* f. ***Hamilton's equations*** *are the differential equations on* P *given by*

$$\dot{z} = X_f(z). \tag{2.1.6}$$

If (P, Ω) *is a symplectic manifold, define the **Poisson bracket operation*** $\{\cdot, \cdot\} : \mathcal{F}(P) \times \mathcal{F}(P) \to \mathcal{F}(P)$ *by*

$$\{f, g\} = \Omega(X_f, X_g). \tag{2.1.7}$$

The construction (2.1.7) makes $(P, \{\,,\})$ into a Poisson manifold. In other words,

Proposition 2.1 *Every symplectic manifold is Poisson.*

The converse is not true; for example the zero bracket makes any manifold Poisson. In §2.4 we shall see some non-trivial examples, such as Lie-Poisson structures on duals of Lie algebras.

Hamiltonian vector fields are defined on Poisson manifolds as follows.

Definition 2.5 *Let* $(P, \{\,,\})$ *be a Poisson manifold and let* $f \in \mathcal{F}(P)$. *Define* X_f *to be the unique vector field on* P *satisfying*

$$X_f[k] := \langle \mathrm{d}k, X_f \rangle = \{k, f\} \quad \text{for all} \quad k \quad \text{in} \quad \mathcal{F}(P).$$

We call X_f *the* **Hamiltonian vector field** *of* f.

A check of the definitions shows that in the symplectic case, the definitions 2.4 and 2.6 of Hamiltonian vector fields coincide. If $(P, \{\,,\})$ is a Poisson manifold, there are therefore three equivalent ways to write Hamilton's equations for $H \in \mathcal{F}(P)$:

    i $\dot{z} = X_H(z)$

    ii $\dot{f} = \mathrm{d}f(z) \cdot X_H(z)$ for all f in $\mathcal{F}(P)$, and

    iii $\dot{f} = \{f, H\}$ for all f in $\mathcal{F}(P)$.

2.2 The Flow of a Hamiltonian Vector Field

Hamilton's equations described in the abstract setting of the last section are very general. They include not only what one normally thinks of as Hamilton's canonical equations in classical mechanics, but Schrödinger's equation in quantum mechanics as well. Despite this generality, the theory has a rich structure.

Let $H \in \mathcal{F}(P)$ where P is a Poisson manifold. Let φ_t be the flow of Hamilton's equations; thus, $\varphi_t(z)$ is the integral curve of $\dot{z} = X_H(z)$ starting at z. (If the flow is not complete, restrict attention to its domain of definition.) There are two basic facts about Hamiltonian flows (ignoring functional analytic technicalities in the infinite dimensional case — see Chernoff and Marsden [1974]).

Proposition 2.2 *The following hold for Hamiltonian systems on Poisson manifolds:*

    i   *each* φ_t *is a Poisson map*

    ii $H \circ \varphi_t = H$ *(conservation of energy).*

The first part of this proposition is true even if H is a time dependent Hamiltonian, a fact that is important to notice. The second part is true only when H is independent of time.

2.3 Cotangent Bundles

Let Q be a given manifold (usually the configuration space of a mechanical system) and T^*Q be its cotangent bundle. Coordinates q^i on Q induce coordinates (q^i, p_j) on T^*Q, called the **canonical cotangent coordinates** of T^*Q.

Proposition 2.3 *There is a unique 1-form* Θ *on* T^*Q *such that in any choice of canonical cotangent coordinates,*

$$\Theta = p_i dq^i; \qquad (2.3.1)$$

Θ *is called the* **canonical 1-form***. We define the* **canonical 2-form** Ω *by*

$$\Omega = -d\Theta = dq^i \wedge dp_i \quad \text{(a sum on i is understood)}. \qquad (2.3.2)$$

In infinite dimensions one needs to use an intrinsic definition of Θ, and there are many such; one of these is the identity $\beta^*\Theta = \beta$ for $\beta : Q \to T^*Q$ any one form.

Proposition 2.4 (T^*Q, Ω) *is a symplectic manifold.*

In canonical coordinates the Poisson brackets on T^*Q have the classical form

$$\{f, g\} = \frac{\partial f}{\partial q^i} \frac{\partial g}{\partial p_i} - \frac{\partial g}{\partial q^i} \frac{\partial f}{\partial p_i} \qquad (2.3.3)$$

where summation on repeated indices is understood.

Theorem 2.1 Darboux' Theorem *Every symplectic manifold locally looks like* T^*Q*; in other words, on every finite dimensional symplectic manifold, there are local coordinates in which* Ω *has the form (2.3.2).*

(See Marsden [1981] and Olver [1988] for a discussion of the infinite dimensional case.)

Hamilton's equations in these canonical coordinates take on the classical form

$$\dot{q}^i = \frac{\partial H}{\partial p_i}$$

$$\dot{p}_i = -\frac{\partial H}{\partial q^i}$$

as one can readily check.

The local structure of Poisson manifolds is more complex than the symplectic case. However, Kirillov has shown that every Poisson manifold is the union of **symplectic leaves**; to compute the bracket of two functions in P, one does it **leaf-wise**. In other words, to calculate the bracket of f and g at $z \in P$, select the symplectic leaf S_z through z, and evaluate the bracket of $f|S_z$ and $g|S_z$ at z. We shall see a specific case of this picture shortly.

2.4   Lagrangian Mechanics

Let Q be a manifold and TQ its tangent bundle. Coordinates q^i on Q induce coordinates $(q^i, \dot{q}^i)$ on TQ, called **tangent coordinates**. A mapping $L : TQ \to \mathbb{R}$ is called a **Lagrangian**. Typically we choose L to be $L = K - V$ where $K(v) = \frac{1}{2}\langle v, v \rangle$ is the **kinetic energy** associated to a given Riemannian metric and where $V : Q \to \mathbb{R}$ is the **potential energy**.

Definition 2.6 *The principle of critical action singles out particular curves $q(t)$ by the condition*

$$\delta \int_b^a L(q(t), \dot{q}(t))dt = 0, \tag{2.4.1}$$

where the variation is over smooth curves in Q with fixed endpoints.

It is interesting to note that (2.4.1) is unchanged if we replace the integrand by $L(q, \dot{q}) - \frac{d}{dt}S(q, t)$ for any function $S(q, t)$. This reflects the **gauge invariance** of classical mechanics and is closely related to Hamilton-Jacobi theory. We shall return to this point in Chapter 9.

If one prefers, the action principle states that the map I defined by $I(q(\cdot)) = \int_a^b L(q(t), \dot{q}(t))dt$ from the space of curves with prescribed endpoints in Q to $\mathbb{R}$ has a critical point at the curve in question. In any case, a basic and elementary result of the calculus of variations is:

Proposition 2.5 *The principle of critical action for a curve $q(t)$ is equivalent to the condition that that curve satisfy the **Euler-Lagrange equations***

$$\frac{d}{dt}\frac{\partial L}{\partial \dot{q}^i} = \frac{\partial L}{\partial q^i}. \tag{2.4.2}$$

Definition 2.7 *Let L be a Lagrangian on TQ and let $\mathbb{F}L : TQ \to T^*Q$ be defined (in coordinates) by*

$$(q^i, \dot{q}^j) \mapsto (q^i, p_j)$$

*where $p_j = \partial L/\partial \dot{q}^j$ denotes the fiber derivative; $\mathbb{F}L$ is called a **Legendre transformation**. (Intrinsically, $\mathbb{F}L$ differentiates L in the fiber direction.)*

A Lagrangian L is called **hyperregular** if $\mathbb{F}L$ is a diffeomorphism. If L is a hyperregular Lagrangian, we define the corresponding **Hamiltonian** by

$$H(q^i, p_j) = p_i \dot{q}^i - L.$$

One checks that the Euler-Lagrange equations for L are equivalent to Hamilton's equations for H.

In a relativistic context one finds that the two conditions $p_j = \partial L/\partial \dot{q}^j$ and $H = p_i \dot{q}^i - L$ fit together as the spatial and temporal components of a single object. Suffice it to say that the formalism developed here is useful in the context of relativistic fields.

2.5   Lie-Poisson Structures

Not every Poisson manifold is symplectic. For example, a large class of non-symplectic Poisson manifolds is the class of Lie-Poisson manifolds, which we now define. Let G be a Lie group and $\mathfrak{g} = T_e G$ its Lie algebra with $[\,,] : \mathfrak{g} \times \mathfrak{g} \to \mathfrak{g}$ the associated Lie bracket.

Proposition 2.6 *The dual space $\mathfrak{g}^*$ is a Poisson manifold with either of the two brackets*

$$\{f, k\}_{\pm}(\mu) = \pm \left\langle \mu, \left[\frac{\delta f}{\delta \mu}, \frac{\delta k}{\delta \mu}\right] \right\rangle. \qquad (2.5.1)$$

Here $\mathfrak{g}$ is identified with $\mathfrak{g}^{**}$ in the sense that $\delta f/\delta \mu \in \mathfrak{g}$ is defined by $\langle \nu, \delta f/\delta \mu \rangle = \mathbf{D}f(\mu) \cdot \nu$ for $\nu \in \mathfrak{g}^*$, where $\mathbf{D}$ denotes the derivative. (In the infinite dimensional case one needs to worry about the existence of $\delta f/\delta \mu$; naive methods like the Hahn-Banach theorem are not always appropriate!) The notation $\delta f/\delta \mu$ is used to conform to the functional derivative notation in classical field theory. In coordinates, $(\xi^1, \ldots, \xi^m)$ on $\mathfrak{g}$ and corresponding dual coordinates $(\mu_1, \ldots, \mu_m)$ on $\mathfrak{g}^*$, the *Lie-Poisson bracket* (2.5.1) is

$$\{f, k\}_{\pm}(\mu) = \pm \mu_a C^a_{bc} \frac{\partial f}{\partial \mu_b} \frac{\partial k}{\partial \mu_c}; \qquad (2.5.2)$$

here C^a_{bc} are the *structure constants* of $\mathfrak{g}$ defined by $[e_a, e_b] = C^c_{ab} e_c$, where $(e_1, \ldots, e_m)$ is the coordinate basis of $\mathfrak{g}$ and where, for $\xi \in \mathfrak{g}$, we write $\xi = \xi^a e_a$, and for $\mu \in \mathfrak{g}^*, \mu = \mu_a e^a$, where $(e^1, \ldots, e^m)$ is the dual basis. The Lie-Poisson reduction of cotangent bundle actions shows which sign has to be taken in the problems under consideration, as we shall see next. Formula (2.5.2) appears explicitly in Lie [1890], §75.

Lie-Poisson reduction can be summarized as follows. Let

$$\lambda : T^*G \to \mathfrak{g}^* \quad \text{be defined by} \quad p_g \mapsto (T_e L_g)^* p_g \in T_e^* G \cong \mathfrak{g}^* \qquad (2.5.3)$$

and

$$\rho : T^*G \to \mathfrak{g}^* \quad \text{be defined by} \quad p_g \mapsto (T_e R_g)^* p_g \in T_e^* G \cong \mathfrak{g}^*. \qquad (2.5.4)$$

Then λ is a Poisson map if one takes the $-$ Lie-Poisson structure on $\mathfrak{g}^*$ and ρ is a Poisson map if one takes the $+$ Lie-Poisson structure on $\mathfrak{g}^*$.

Every left invariant Hamiltonian and Hamiltonian vector field is mapped by λ to a Hamiltonian and Hamiltonian vector field on $\mathfrak{g}^*$. There is a similar statement for right invariant systems on T^*G. One says that the original system on T^*G has been **reduced** to $\mathfrak{g}^*$. The reason λ and ρ are both Poisson maps is perhaps best understood by observing that they are both equivariant momentum maps generated by the action of G on itself by right and left translations, respectively. We turn to this topic in §2.6.

2.6   The Rigid Body

Let $\mathbb{I}$ be a 3×3 positive definite symmetric tensor (the **moment of inertia tensor**), ω be the **body angular velocity** and $\Pi = \mathbb{I}\omega$ be the **body angular momentum**. The **Euler equations** of motion for rigid body dynamics are given by

$$\dot{\Pi} = \Pi \times \omega, \tag{2.6.1}$$

which for $\mathbb{I} = \mathrm{diag}\,(I_1, I_2, I_3)$ is diagonal, agree with those in Lecture 1. Euler's equations are Hamiltonian relative to a Lie-Poisson structure. To see this, take $G = SO(3)$ to be the configuration space of the rigid body for movement about a fixed point. Then $\mathfrak{g} \cong (\mathbb{R}^3, \times)$ and we identify $\mathfrak{g} \cong \mathfrak{g}^*$. The corresponding Lie-Poisson structure on $\mathbb{R}^3$ is given by

$$\{f, k\}(\Pi) = -\Pi \cdot (\nabla f \times \nabla k). \tag{2.6.2}$$

For the rigid body one chooses the minus sign in the Lie-Poisson bracket. We shall see how to *derive* (rather than guess) the expression (2.6.2) in §2.8.

Starting with the kinetic energy Hamiltonian derived in Chapter 1, we directly obtain the formula $H(\Pi) = \frac{1}{2}\Pi \cdot (\mathbb{I}^{-1}\Pi)$, the kinetic energy of the rigid body. One verifies from the chain rule and properties of the triple product that:

Proposition 2.7 *Euler's equations are equivalent to the following equation for all* $f \in \mathcal{F}(\mathbb{R}^3)$:

$$\dot{f} = \{f, H\}. \tag{2.6.3}$$

Definition 2.8 *Let* $(P, \{\,,\})$ *be a Poisson manifold. A function* $C \in \mathcal{F}(P)$ *satisfying*

$$\{C, f\} = 0 \quad \text{for all} \quad f \quad \text{in} \quad \mathcal{F}(P) \tag{2.6.4}$$

*is called a **Casimir function**.*

A crucial difference between symplectic manifolds and Poisson manifolds is this: On symplectic manifolds, the only Casimir functions are clearly the constant functions (assuming P is connected). On the other hand, on Poisson manifolds there is often a large supply of Casimir functions. In the case of the rigid body, every function $C : \mathbb{R}^3 \to \mathbb{R}$ of the form

$$C(\Pi) = \Phi(\|\Pi\|^2) \qquad (2.6.5)$$

where $\Phi : \mathbb{R} \to \mathbb{R}$ is a differentiable function, is a Casimir function, as we noted in Chapter 1. Casimir functions are constants of the motion for *any* Hamiltonian since $\dot{C} = \{C, H\} = 0$ for any H. In particular, for the rigid body, $\|\Pi\|^2$ is a constant of the motion — this is the invariant sphere we saw in Lecture 1.

There is an intimate relation between Casimirs and symmetry generated conserved quantities, or *momentum maps*. We turn to them next.

2.7   Momentum Maps

Let G be a Lie group and P be a Poisson manifold, such that G acts on P by Poisson maps (in this case the action is called a *Poisson action*). Denote the corresponding infinitesimal action of $\mathfrak{g}$ on P by $\xi \mapsto \xi_P$, a map of $\mathfrak{g}$ to $\mathfrak{X}(P)$, the space of vector fields on P. We write the action of $g \in G$ on $z \in P$ as simply gz; the vector field ξ_P is obtained at z by differentiating gz with respect to g in the direction ξ at $g = e$. Explicitly,

$$\xi_P(z) = \frac{d}{d\epsilon}[\exp(\epsilon\xi) \cdot z]\Big|_{\epsilon=0}.$$

Definition 2.9 *A map* $\mathbf{J} : P \to \mathfrak{g}^*$ *is called a* momentum map *if* $X_{\langle \mathbf{J}, \xi \rangle} = \xi_P$ *for each* $\xi \in \mathfrak{g}$, *where* $\langle \mathbf{J}, \xi \rangle(z) = \langle \mathbf{J}(z), \xi \rangle$.

Theorem 2.2 (Noether's Theorem) *If* H *is a* G *invariant Hamiltonian on* P, *then* $\mathbf{J}$ *is conserved on the trajectories of the Hamiltonian vector field* X_H.

Proof Differentiating the invariance condition $H(gz) = H(z)$ with respect to $g \in G$ for fixed $z \in P$, we get $\mathbf{d}H(z) \cdot \xi_P(z) = 0$ and so $\{H, \langle \mathbf{J}, \xi \rangle\} = 0$ which by antisymmetry gives $\mathbf{d}\langle \mathbf{J}, \xi \rangle \cdot X_H = 0$ and so $\langle \mathbf{J}, \xi \rangle$ is conserved on the trajectories of X_H for every ξ in G. ■

Turning to the construction of momentum maps, let Q be a manifold and let G act on Q. This action induces an action of G on T^*Q by cotangent lift — that is, we take the transpose inverse of the tangent lift. The action of G on T^*Q is always symplectic and therefore Poisson.

Theorem 2.3 *A momentum map for a cotangent lifted action is given by*

$$\mathbf{J} : T^*Q \to \mathfrak{g}^* \quad \textit{defined by} \quad \langle \mathbf{J}, \xi \rangle (p_q) = \langle p_q, \xi_Q(q) \rangle. \tag{2.7.1}$$

In canonical coordinates, we write $p_q = (q^i, p_j)$ and define the **action functions** $A^i{}_a$ by $(\xi_Q)^i = A^i{}_a(q)\xi^a$. Then

$$\langle \mathbf{J}, \xi \rangle (p_q) = p_i A^i{}_a(q)\xi^a \tag{2.7.2}$$

and therefore

$$J_a = p_i A^i{}_a(q). \tag{2.7.3}$$

Recall that by differentiating conjugation at the identity one gets the **adjoint action** of G on $\mathfrak{g}$. By taking its dual one gets the **coadjoint action** of G on $\mathfrak{g}^*$.

Proposition 2.8 *The momentum map for cotangent lifted actions is equivariant, i.e., the diagram in Figure 2.7.1 commutes.*

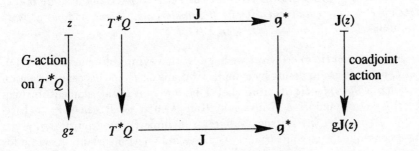

Figure 2.7.1: Equivariance of the momentum map.

Proposition 2.9 *Equivariance implies infinitesimal equivariance, which can be stated as the **classical commutation relations**:*

$$\{\langle \mathbf{J}, \xi \rangle, \langle \mathbf{J}, \eta \rangle\} = \langle \mathbf{J}, [\xi, \eta] \rangle.$$

Proposition 2.10 *If $\mathbf{J}$ is infinitesimally equivariant, then $\mathbf{J} : P \to \mathfrak{g}^*$ is a Poisson map. If $\mathbf{J}$ is generated by a left (respectively right) action then we use the $+$ (respectively $-$) Lie-Poisson structure on $\mathfrak{g}^*$.*

2.8   Reduction

There are three levels of reduction of decreasing generality, that of Poisson reduction, symplectic reduction, and cotangent bundle reduction. Let us first consider Poisson reduction.

For **Poisson reduction** we start with a Poisson manifold P and let the Lie group G act on P by Poisson maps. Assuming P/G is a smooth manifold, endow it with the unique Poisson structure on such that the canonical projection $\pi : P \to P/G$ is a Poisson map. We can specify the Poisson structure on P/G explicitly as follows. For f and $k : P/G \to \mathbb{R}$, let $F = f \circ \pi$ and $K = k \circ \pi$, so F and K are f and k thought of as G-invariant functions on P. Then $\{f, k\}_{P/G}$ is defined by

$$\{f, k\}_{P/G} \circ \pi = \{F, K\}_P. \tag{2.8.1}$$

To show that $\{f, k\}_{P/G}$ is well defined, one has to prove that $\{F, K\}_P$ is G-invariant. This follows from the fact that F and K are G-invariant and the group action of G on P consists of Poisson maps.

For $P = T^*G$ we get a very important special case.

Theorem 2.4 Lie-Poisson Reduction *Let $P = T^*G$ and assume that G acts on P by the cotangent lift of left translations. If one endows $\mathfrak{g}^*$ with the minus Lie-Poisson bracket, then $P/G \cong \mathfrak{g}^*$.*

For **symplectic reduction** we begin with a symplectic manifold (P, Ω). Let G be a Lie group acting by symplectic maps on P; in this case the action is called a **symplectic action**. Let $\mathbf{J}$ be an equivariant momentum map for this action and H a G-invariant Hamiltonian on P. Let $G_\mu = \{g \in G \mid g \cdot \mu = \mu\}$ be the isotropy subgroup (symmetry subgroup) at $\mu \in \mathfrak{g}^*$. As a consequence of equivariance, G_μ leaves $\mathbf{J}^{-1}(\mu)$ invariant. Assume for simplicity that μ is a regular value of $\mathbf{J}$, so that $\mathbf{J}^{-1}(\mu)$ is a smooth manifold (see §2.8 below) and that G_μ acts freely and properly on $\mathbf{J}^{-1}(\mu)$, so that $\mathbf{J}^{-1}(\mu)/G_\mu =: P_\mu$ is a smooth manifold. Let $i_\mu : \mathbf{J}^{-1}(\mu) \to P$ denote the inclusion map and let $\pi_\mu : \mathbf{J}^{-1}(\mu) \to P_\mu$ denote the projection. Note that

$$\dim P_\mu = \dim P - \dim G - \dim G_\mu. \tag{2.8.2}$$

Building on classical work of Jacobi, Liouville, Arnold and Smale, we have

Theorem 2.5 (Reduction Theorem) *(Meyer [1973], Marsden and Weinstein [1974]) There is a unique symplectic structure Ω_μ on P_μ satisfying*

$$i_\mu^* \Omega = \pi_\mu^* \Omega_\mu. \tag{2.8.3}$$

Given a G-invariant Hamiltonian H on P, define the reduced Hamiltonian $H_\mu : P_\mu \to \mathbb{R}$ by $H = H_\mu \circ \pi_\mu$. Then the trajectories of X_H project to those of X_{H_μ}. An important problem is how to reconstruct trajectories of X_H from trajectories of X_{H_μ}. Schematically, we have the situation in Figure 2.8.1.

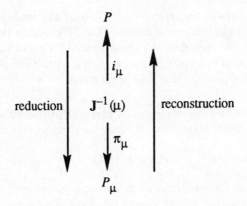

Figure 2.8.1: Reduction to P_μ and reconstruction back to P.

As we shall see later, the reconstruction process is where the holonomy and "geometric phase" ideas enter. In fact, we shall put a connection on the bundle $\pi_\mu : \mathbf{J}^{-1}(\mu) \to P_\mu$ and it is through this process that one encounters the gauge theory point of view of mechanics.

Let $\mathcal{O}_\mu$ denote the coadjoint orbit through μ. As a special case of the symplectic reduction theorem, we get

Corollary 2.1 $(T^*G)_\mu \cong \mathcal{O}_\mu$.

The symplectic structure inherited on $\mathcal{O}_\mu$ is called the *(Lie-Kostant-Kirillov) orbit symplectic structure*. This structure is compatible with the Lie-Poisson structure on $\mathfrak{g}^*$ in the sense that the bracket of two functions on $\mathcal{O}_\mu$ equals that obtained by extending them arbitrarily to $\mathfrak{g}^*$, taking the Lie-Poisson bracket on $\mathfrak{g}^*$ and then restricting to $\mathcal{O}_\mu$.

Example 1 $G = SO(3), \mathfrak{g}^* = so(3)^* \cong \mathbb{R}^3$. In this case the coadjoint action is the usual action of $SO(3)$ on $\mathbb{R}^3$. This is because of the orthogonality of the elements of G. The set of orbits consists of spheres and a single point. The reduction process confirms that all orbits are symplectic manifolds. One calculates that the symplectic structure on the spheres is a multiple of the area element.

Example 2 *Jacobi-Liouville theorem* Let $G = \mathbb{T}^k$ be the k-torus and assume G acts on a symplectic manifold P. In this case the components of $\mathbf{J}$ are in involution and $\dim P_\mu = \dim P - 2k$, so $2k$ variables are eliminated. As we shall see, reconstruction allows one to reassemble the solution trajectories on P by quadratures in this abelian case.

Example 3 *Jacobi-Deprit elimination of the node*) Let $G = SO(3)$ act on P. In the classical case of Jacobi, $P = T^*\mathbb{R}^3$ and in the generalization of Deprit [1983] one considers the phase space of n particles in $\mathbb{R}^3$. We just point out here that the reduced space P_μ has dimension $\dim P - 3 - 1 = \dim P - 4$ since $G_\mu = S^1$ (if $\mu \neq 0$) in this case. ◆

The *orbit reduction theorem* of Marle [1976] and Kazhdan, Kostant and Sternberg [1978] states that P_μ may be alternatively constructed as

$$P_\mathcal{O} = \mathbf{J}^{-1}(\mathcal{O})/G, \tag{2.8.4}$$

where $\mathcal{O} \subset \mathfrak{g}^*$ is the coadjoint orbit through μ. As above we assume we are away from singular points (see §2.8 below). The spaces P_μ and $P_\mathcal{O}$ are isomorphic by using the inclusion map $l_\mu : \mathbf{J}^{-1}(\mu) \to \mathbf{J}^{-1}(\mathcal{O})$ and taking equivalence classes to induce a symplectic isomorphism $L_\mu : P_\mu \to P_\mathcal{O}$. The symplectic structure $\Omega_\mathcal{O}$ on $P_\mathcal{O}$ is determined by

$$j_\mathcal{O}^* \Omega = \pi_\mathcal{O}^* \Omega_\mathcal{O} + \mathbf{J}_\mathcal{O}^* \omega_\mathcal{O} \tag{2.8.5}$$

where $j_\mathcal{O} : \mathbf{J}^{-1}(\mathcal{O}) \to P$ is the inclusion, $\pi_\mathcal{O} : \mathbf{J}^{-1}(\mathcal{O}) \to P_\mathcal{O}$ is the projection, and where $\mathbf{J}_\mathcal{O} = \mathbf{J}|\mathbf{J}^{-1}(\mathcal{O}) : \mathbf{J}^{-1}(\mathcal{O}) \to \mathcal{O}$ and $\omega_\mathcal{O}$ is the orbit symplectic form. In terms of the Poisson structure, $\mathbf{J}^{-1}(\mathcal{O})/G$ has the bracket structure inherited from P/G; in fact, $\mathbf{J}^{-1}(\mathcal{O})/G$ *is a symplectic leaf* in P/G. Thus, we get the picture in Figure 2.8.2.

Kirillov has shown that *every* Poisson manifold P is the union of symplectic leaves, although the preceding construction explicitly realizes these symplectic leaves in this case by the reduction construction. A special case is the foliation of the dual $\mathfrak{g}^*$ of any Lie algebra $\mathfrak{g}$ into its symplectic leaves, namely the coadjoint orbits. For example $SO(3)$ is the union of spheres plus the origin, each of which is a symplectic manifold. Notice that the drop in dimension from $T^*SO(3)$ to $\mathcal{O}$ is from 6 to 2, a drop of 4, as in general $SO(3)$ reduction. An exception is the singular point, the origin, where the drop in dimension is larger. We turn to these singular points next.

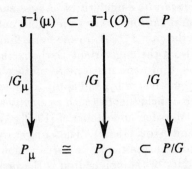

Figure 2.8.2: Orbit reduction gives another realization of P_μ.

2.9 Singularities and Symmetry

Proposition 2.11 *Let (P, Ω) be a symplectic manifold, let G act on P by Poisson mappings, and let $\mathbf{J} : P \to \mathfrak{g}^*$ be a momentum map for this action ($\mathbf{J}$ need not be equivariant). Let G_z denote the symmetry group of $z \in P$ defined by $G_z = \{g \in G \mid gz = z\}$ and let $\mathfrak{g}_z$ be its Lie algebra, so $\mathfrak{g}_z = \{\zeta \in \mathfrak{g} \mid \zeta_P(z) = 0\}$. Then z is a regular value of $\mathbf{J}$ if and only if $\mathfrak{g}_z$ is trivial; i.e., $\mathfrak{g}_z = \{0\}$, or G_z is discrete.*

Proof The point z is regular when the range of the linear map $\mathbf{DJ}(z)$ is all of $\mathfrak{g}^*$. However, $\zeta \in \mathfrak{g}$ is orthogonal to the range (in the sense of the $\mathfrak{g}, \mathfrak{g}^*$ pairing) if and only if for all $v \in T_z P$,

$$\langle \zeta, \mathbf{DJ}(z) \cdot v \rangle = 0$$

i.e.,

$$\mathbf{d}\langle \mathbf{J}, \zeta \rangle(z) \cdot v = 0$$

or

$$\Omega(X_{\langle \mathbf{J}, \zeta \rangle}(z), v) = 0$$

or

$$\Omega(\zeta_P(z), v) = 0.$$

As Ω is nondegenerate, ζ is orthogonal to the range iff $\zeta_P(z) = 0$. ∎

The above proposition is due to Smale [1970]. It is the starting point of a large literature on singularities in the momentum map. Arms, Marsden and Moncrief [1981] show, under some reasonable hypotheses, that the level sets $\mathbf{J}^{-1}(0)$ have *quadratic* singularities. As we shall see in the next chapter, there is a general *shifting construction* that enables one to effectively reduce $\mathbf{J}^{-1}(\mu)$ to the case $\mathbf{J}^{-1}(0)$. In the finite dimensional case, this result can be deduced from the equivariant Darboux theorem, but in the infinite dimensional case, things are much more subtle. In fact, the infinite dimensional results were motivated by, and apply to, the singularities in the solution space of relativistic field theories such as gravity and the Yang-Mills equations (see Fischer, Marsden and Moncrief [1980], Arms, Marsden and Moncrief [1981, 1982] and Arms [1981]). The *convexity theorem* states that the image of the momentum map of a torus action is a convex polyhedron in $\mathfrak{g}^*$; the boundary of the polyhedron is the image of the singular (symmetric) points in P; the more symmetric the point, the more singular the boundary point. These results are due to Atiyah [1982] and Guillemin and Sternberg [1984] based on earlier convexity results of Kostant and the Shur-Horne theorem on eigenvalues of symmetric matrices. The literature on these topics and its relation to other areas of mathematics is vast. See, for example, Goldman and Millison [1990], Sjamaar [1990], Bloch, Flaschka and Ratiu [1990] and Lu and Ratiu [1991].

2.10   A Particle in a Magnetic Field

During cotangent bundle reduction considered in the next chapter, we shall have to add terms to the symplectic form called "magnetic terms". To explain this terminology, we consider a particle in a magnetic field.

Let B be a closed two-form on $\mathbb{R}^3$ and $\mathbf{B} = B_x\mathbf{i} + B_y\mathbf{j} + B_z\mathbf{k}$ the associated divergence free vector field, *i.e.*, $\mathbf{i_B}(dx \wedge dy \wedge dz) = B$, or

$$B = B_x dy \wedge dz - B_y dx \wedge dz + B_z dx \wedge dy.$$

Thinking of $\mathbf{B}$ as a magnetic field, the equations of motion for a particle with charge e and mass m are given by the *Lorentz force law*:

$$m\frac{d\mathbf{v}}{dt} = \frac{e}{c}\mathbf{v} \times \mathbf{B} \tag{2.10.1}$$

where $\mathbf{v} = (\dot{x}, \dot{y}, \dot{z})$. On $\mathbb{R}^3 \times \mathbb{R}^3$ *i.e.*, on $(\mathbf{x}, \mathbf{v})$-space, consider the symplectic form

$$\Omega_B = m(dx \wedge d\dot{x} + dy \wedge d\dot{y} + dz \wedge d\dot{z}) - \frac{e}{c}B. \tag{2.10.2}$$

For the Hamiltonian, take the kinetic energy:

$$H = \frac{m}{2}(\dot{x}^2 + \dot{y}^2 + \dot{z}^2) \tag{2.10.3}$$

writing $X_H(u, v, w) = (u, v, w, (\dot{u}, \dot{v}, \dot{w}))$, the condition defining X_H, namely $\mathbf{i}_{X_H}\Omega_B = \mathbf{d}H$ is

$$m(ud\dot{x} - \dot{u}dx + vd\dot{y} - \dot{v}dy + wd\dot{z} - \dot{w}dz)$$
$$- \frac{e}{c}[B_x vdz - B_x wdy - B_y udz + B - ywdx + B_z udy - B_z vdx]$$
$$= m(\dot{x}d\dot{x} + \dot{y}d\dot{y} + \dot{z}d\dot{z}) \tag{2.10.4}$$

which is equivalent to $u = \dot{x}, v = \dot{y}, w = \dot{z}, m\dot{u} = e(B_z v - B_y w)/c, m\dot{v} = e(B_x w - B_z u)/c$, and $m\dot{w} = e(B_y u - B_x v)/c$, *i.e.*, to

$$m\ddot{x} = \frac{e}{c}(B_z \dot{y} - B_y \dot{z})$$
$$m\ddot{y} = \frac{e}{c}(B_x \dot{z} - B_z \dot{x}) \tag{2.10.5}$$
$$m\ddot{z} = \frac{e}{c}(B_y \dot{y} - B_x \dot{z})$$

which is the same as (2.10.1). Thus the *equations of motion for a particle in a magnetic field are Hamiltonian, with energy equal to the kinetic energy and with the symplectic form* Ω_B.

If $B = \mathbf{d}A$; *i.e.*, $\mathbf{B} = \nabla \times \mathbf{A}$, where A is a one-form and $\mathbf{A}$ is the associated vector field, then the map $(\mathbf{x}, \mathbf{v}) \mapsto (\mathbf{x}, \mathbf{p})$ where $\mathbf{p} = m\mathbf{v} + e\mathbf{A}/c$ pulls back the *canonical* form to Ω_B, as is easily checked. *Thus, equations (2.10.1) are also Hamiltonian relative to the canonical bracket on* $(\mathbf{x}, \mathbf{p})$-*space with the Hamiltonian*

$$H_{\mathbf{A}} = \frac{1}{2m}\|\mathbf{p} - \frac{e}{c}\mathbf{A}\|^2. \tag{2.10.6}$$

Even in Euclidean space, not every magnetic field can be written as $\mathbf{B} = \nabla \times \mathbf{A}$. For example, the field of a magnetic monopole of strength $g \neq 0$, namely

$$\mathbf{B}(\mathbf{r}) = g\frac{\mathbf{r}}{\|\mathbf{r}\|^3} \tag{2.10.7}$$

cannot be written this way since the flux of $\mathbf{B}$ through the unit sphere is $4\pi g$, yet Stokes' theorem applied to the two hemispheres would give zero. Thus, one might think that the Hamiltonian formulation involving only B (*i.e.*, using Ω_B and H) is preferable. However, one can recover the magnetic potential A by regarding A as a connection on a nontrivial bundle over $\mathbb{R}^3 \backslash \{0\}$. The bundle over the sphere S^2 is in fact the same *Hopf fibration* $S^3 \to S^2$ that we encountered in §1.6. This same construction can be carried out using reduction. For a readable account of some aspects of this situation, see Yang [1980]. For an interesting example of Weinstein in which this monopole comes up, see Marsden [1981], p. 34.

When one studies the motion of a colored (rather than a charged) particle in a Yang-Mills field, one finds a beautiful generalization of this construction and related ideas using the theory of principal bundles; see Sternberg [1977], Weinstein [1978] and Montgomery [1985]. In the study of centrifugal and Coriolis forces one discovers some structures analogous to those here. We shall return to particles in Yang-Mills fields in our discussion of optimal control in Chapter 7.

Chapter 3

Cotangent Bundle Reduction

In this chapter we discuss the cotangent bundle reduction theorem. Versions of this are already given in Smale [1970], but primarily for the abelian case. This was amplified in the work of Satzer [1977] and motivated by this, was extended to the nonabelian case in Abraham and Marsden [1978]. An important formulation of this was given by Kummer [1981] in terms of connections. Building on this, the "bundle picture" was developed by Montgomery, Marsden and Ratiu [1984] and Montgomery [1986].

From the symplectic viewpoint, the principal result is that the reduction of a cotangent bundle T^*Q at $\mu \in \mathfrak{g}^*$ is a bundle over $T^*(Q/G)$ with fiber the coadjoint orbit through μ. Here, $S = Q/G$ is called *shape space*. From the Poisson viewpoint, this reads: $(T^*Q)/G$ is a $\mathfrak{g}^*$-bundle over $T^*(Q/G)$, or a Lie-Poisson bundle over the cotangent bundle of shape space. We describe the geometry of this reduction using the mechanical connection and explicate the reduced symplectic structure and the reduced Hamiltonian for simple mechanical systems.

3.1   Mechanical G-systems

By a *symplectic* (resp. *Poisson*) *G-system* we mean a symplectic (resp. Poisson) manifold (P, Ω) together with the symplectic action of a Lie group G on P, an *equivariant* momentum map $\mathbf{J} : P \to \mathfrak{g}^*$ and a G-invariant Hamiltonian $H : P \to \mathbb{R}$.

Following terminology of Smale [1970], we refer to the following special case of a symplectic G-system as a *simple mechanical G-system*. We choose $P = T^*Q$, assume there is a Riemannian metric $\langle\langle\,,\rangle\rangle$ on Q, that G acts on Q by isometries (and so G acts on T^*Q by cotangent lifts) and that

$$H(q,p) = \frac{1}{2}\|p\|_q^2 + V(q), \tag{3.1.1}$$

where $\|\cdot\|_q$ is the norm induced on T_q^*Q, and where V is a G-invariant potential.

We abuse notation slightly and write either $z = (q,p)$ or $z = p_q$ for a covector based at $q \in Q$ and we shall also let $\|\cdot\|_q$ denote the norm on T_qQ. Points in T_qQ shall be denoted v_q or (q,v) and the pairing between T_q^*Q and T_qQ is simply written

$$\langle p_q, v_q \rangle, \quad \langle p, v \rangle \quad \text{or} \quad \langle (q,p),(q,v)\rangle. \tag{3.1.2}$$

Other natural pairings between spaces and their duals are also denoted $\langle\,,\rangle$. The precise meaning will always be clear from the context.

Unless mentioned to the contrary, the momentum map for simple mechanical G-systems will be the standard one:

$$\mathbf{J} : T^*Q \to \mathfrak{g}^*, \quad \text{where} \quad \langle \mathbf{J}(q,p),\xi \rangle = \langle p, \xi_Q(q)\rangle \tag{3.1.3}$$

where ξ_Q denotes the infinitesimal generator of ξ on Q.

For the convenience of the reader we will write most formulas in both coordinate free notation and in coordinates using standard tensorial conventions. As in Chapter 2, coordinate indices for P are denoted z^I, z^J, etc., on Q by q^i, q^j, etc., and on $\mathfrak{g}$ (relative to a vector space basis of $\mathfrak{g}$) by ξ^a, ξ^b, etc.. For instance, $\dot{z} = X_H(z)$ can be written

$$\dot{z}^I = X_H^I(z^J) \quad \text{or} \quad \dot{z}^I = \{z^I, H\}. \tag{3.1.4}$$

Recall from Chapter 2 that the **Poisson tensor** is defined by $B(\mathbf{d}H) = X_H$, or equivalently, $B^{IJ}(z) = \{z^I, z^J\}$, so (3.1.4) reads

$$X_H^I = B^{IJ}\frac{\partial H}{\partial z^J} \quad \text{or} \quad \dot{z}^I = B^{IJ}(z)\frac{\partial H}{\partial z^J}(z). \tag{3.1.5}$$

Equation (3.1.1) reads

$$H(q,p) = \frac{1}{2}g^{ij}p_ip_j + V(q) \tag{3.1.6}$$

and (3.1.2) reads

$$\langle p, v \rangle = p_iv^i, \tag{3.1.7}$$

while (3.1.3) reads

$$J_a(q,p) = p_i A^i{}_a(q) \tag{3.1.8}$$

where the ***action tensor*** $A^i{}_a$ is defined, as in Chapter 2, by

$$[\xi_Q(q)]^i = A^i{}_a(q)\xi^a. \qquad (3.1.9)$$

The Legendre transformation is denoted $\mathbb{F}L : TQ \to T^*Q$ and in the case of simple mechanical systems is simply the metric tensor regarded as a map from vectors to covectors; in coordinates,

$$\mathbb{F}L(q,v) = (q,p), \quad \text{where} \quad p_i = g_{ij}v^j. \qquad (3.1.10)$$

Example 1 *The spherical pendulum* Here $Q = S^2$, the sphere on which the bob moves, the metric is the standard one, the potential is the gravitational potential and $G = S^1$ acts on S^2 by rotations about the vertical axis. The momentum map is simply the angular momentum about the z-axis. We will work this out in more detail below. ◆

Example 2 *The double spherical pendulum* Here the configuration space is $Q = S^2 \times S^2$, the group is S^1 acting by simultaneous rotation about the z-axis and again the momentum map is the total angular momentum about the z-axis. Again, we will see more detail below. ◆

Example 3 *Coupled rigid bodies* Here we have two rigid bodies in $\mathbb{R}^3$ coupled by a ball in socket joint. We choose $Q = \mathbb{R}^3 \times SO(3) \times SO(3)$ describing the joint position and the attitude of each of the rigid bodies relative to a reference configuration $\mathcal{B}$. The Hamiltonian is the kinetic energy, which defines a metric on Q. Here $G = SE(3)$ which acts on the left in the obvious way by transforming the positions of the particles $x_i = A_i X + w, i = 1, 2$ by Euclidean motions. The momentum map is the total linear and angular momentum. If the bodies have additional material symmetry, G is correspondingly enlarged; *cf.* Patrick [1989]. See Figure 3.1.1. ◆

Example 4 *Ideal fluids* For an ideal fluid moving in a container represented by a region $\Omega \subset \mathbb{R}^3$, the configuration space is $Q = \text{Diff}_{\text{vol}}(\Omega)$, the volume preserving diffeomorphisms of Ω to itself. Here $G = Q$ acts on itself on the right and H is the total kinetic energy of the fluid. Here Lie-Poisson reduction is relevant. We refer to Ebin and Marsden [1970] for the relevant functional analytic technicalities. We also refer to Marsden and Weinstein [1982, 1983] for more information and corresponding ideas for plasma physics, and to Marsden and Hughes [1983] and Simo, Marsden and Krishnaprasad [1988] for elasticity. ◆

The examples we will focus on in these lectures are the spherical pendula and the classical water molecule. Let us pause to give some details for the later example.

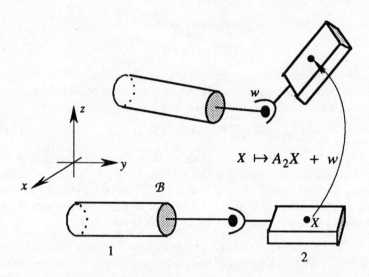

Figure 3.1.1: The configuration space for the dynamics of two coupled rigid bodies.

3.2 The Classical Water Molecule

We started discussing this example in Chapter 1. The "primitive" configuration space of this system is $Q = \mathbb{R}^3 \times \mathbb{R}^3 \times \mathbb{R}^3$, whose points are triples, denoted $(\mathbf{R}, \mathbf{r}_1, \mathbf{r}_2)$, giving the positions of the three masses relative to an inertial frame. For simplicity, we assume nothing singular happens during collisions, as this aspect of the system is not of concern to the present discussion.

The Lagrangian is given on TQ by

$$L(\mathbf{R}, \mathbf{r}_1, \mathbf{r}_2, \dot{\mathbf{R}}, \dot{\mathbf{r}}_1, \dot{\mathbf{r}}_2) = \frac{1}{2}\{M\|\dot{\mathbf{R}}\|^2 + m\|\dot{\mathbf{r}}_1\|^2 + m\|\dot{\mathbf{r}}_2\|^2\} - V(\mathbf{R}, \mathbf{r}_1, \mathbf{r}_2)$$

$$(3.2.1)$$

where V is a potential.

The Legendre transformation gives the Hamiltonian system on $Z = T^*Q$ with

$$H(\mathbf{R}, \mathbf{r}_1, \mathbf{r}_2, \mathbf{P}, \mathbf{p}_1, \mathbf{p}_2) = \frac{\|\mathbf{P}\|^2}{2M} + \frac{\|\mathbf{p}_1\|^2}{2m} + \frac{\|\mathbf{p}_2\|^2}{2m} + V. \qquad (3.2.2)$$

The Euclidean group acts on this system by simultaneous translation and rotation of the three component positions and momenta. In addition, there is a discrete symmetry closely related to interchanging $\mathbf{r}_1$ and $\mathbf{r}_2$ (and, simultaneously, $\mathbf{p}_1$ and $\mathbf{p}_2$). Accordingly, we will assume that V is symmetric in its second two arguments as well as being Euclidean-invariant.

The translation group $\mathbb{R}^3$ acts on Z by

$$\mathbf{a} \cdot (\mathbf{R}, \mathbf{r}_1, \mathbf{r}_2, \mathbf{P}, \mathbf{p}_1, \mathbf{p}_2) = (\mathbf{R} + \mathbf{a}, \mathbf{r}_1 + \mathbf{a}, \mathbf{r}_2 + \mathbf{a}, \mathbf{P}, \mathbf{p}_1, \mathbf{p}_2)$$

and the corresponding momentum map $\mathbf{j} : Z \to \mathbb{R}^3$ is the total linear momentum

$$\mathbf{j}(\mathbf{R}, \mathbf{r}_1, \mathbf{r}_2, \mathbf{P}, \mathbf{p}_1, \mathbf{p}_2) = \mathbf{P} + \mathbf{p}_1 + \mathbf{p}_2. \qquad (3.2.3)$$

Reduction to center of mass coordinates entails that we set $\mathbf{j}$ equal to a constant and quotient by translations. To coordinatize the reduced space (and bring relevant metric tensors into diagonal form), we shall use *Jacobi-Bertrand-Haretu coordinates*, namely

$$\mathbf{r} = \mathbf{r}_2 - \mathbf{r}_1 \quad \text{and} \quad \mathbf{s} = \mathbf{R} - \frac{1}{2}(\mathbf{r}_1 + \mathbf{r}_2) \qquad (3.2.4)$$

as in Figure 3.2.1.

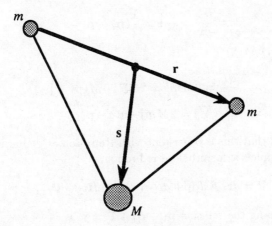

Figure 3.2.1: Jacobi-Bertrand-Haretu coordinates.

To determine the correct momenta conjugate to these variables, we go back to the Lagrangian and write it in the variables $\mathbf{r}, \mathbf{s}, \dot{\mathbf{r}}, \dot{\mathbf{s}}$, using the relations

$$\dot{\mathbf{r}} = \dot{\mathbf{r}}_2 - \dot{\mathbf{r}}_1, \quad \dot{\mathbf{s}} = \dot{\mathbf{R}} - \frac{1}{2}(\dot{\mathbf{r}}_1 + \dot{\mathbf{r}}_2)$$

and

$$\mathbf{j} = m(\dot{\mathbf{r}}_1 + \dot{\mathbf{r}}_2) + M\dot{\mathbf{R}},$$

which give

$$\dot{\mathbf{r}}_1 = \dot{\mathbf{R}} - \dot{\mathbf{s}} - \frac{1}{2}\dot{\mathbf{r}}$$

$$\dot{\mathbf{r}}_2 = \dot{\mathbf{R}} - \dot{\mathbf{s}} + \frac{1}{2}\dot{\mathbf{r}}$$

$$\dot{\mathbf{R}} = \frac{1}{\mathcal{M}}(\mathbf{j} + 2m\dot{\mathbf{s}})$$

where $\mathcal{M} = M + 2m$ is the total mass. Substituting into (3.2.1) and simplifying gives

$$L = \frac{\|\mathbf{j}\|^2}{2\mathcal{M}} + \frac{m}{4}\|\dot{\mathbf{r}}\|^2 + M\overline{m}\|\dot{\mathbf{s}}\|^2 - V \tag{3.2.5}$$

where $\overline{m} = m/\mathcal{M}$ and where one expresses the potential V in terms of $\mathbf{r}$ and $\mathbf{s}$ (if rotationally invariant, it depends on $\|\mathbf{r}\|, \|\mathbf{s}\|$ and $\mathbf{r} \cdot \mathbf{s}$ alone).

A crucial feature of the above choice is that *the Riemannian metric (mass matrix) associated to L is diagonal* as (3.2.5) shows. The conjugate momenta are

$$\pi = \frac{\partial L}{\partial \dot{\mathbf{r}}} = \frac{m}{2}\dot{\mathbf{r}} = \frac{1}{2}(\mathbf{p}_2 - \mathbf{p}_1) \tag{3.2.6}$$

and, with $\overline{M} = M/\mathcal{M}$,

$$\sigma = \frac{\partial L}{\partial \dot{\mathbf{s}}} = 2M\overline{m}\dot{\mathbf{s}} = 2\overline{m}\mathbf{P} - \overline{M}(\mathbf{p}_1 + \mathbf{p}_2) \tag{3.2.7}$$

$$= \mathbf{P} - \overline{M}\mathbf{j} = 2M\overline{m}\mathbf{j} - \mathbf{p}_1 - \mathbf{p}_2. \tag{3.2.8}$$

We next check that this is consistent with Hamiltonian reduction. First, note that the symplectic form before reduction,

$$\Omega = d\mathbf{r}_1 \wedge d\mathbf{p}_1 + d\mathbf{r}_2 \wedge d\mathbf{p}_2 + d\mathbf{R} \wedge d\mathbf{P}$$

can be written, using the relation $\mathbf{p}_1 + \mathbf{p}_2 + \mathbf{P} = \mathbf{j}$, as

$$\Omega = d\mathbf{r} \wedge d\pi + d\mathbf{s} \wedge d\sigma + d\bar{\mathbf{r}} \wedge d\mathbf{j} \tag{3.2.9}$$

where $\bar{\mathbf{r}} = \frac{1}{2}(\mathbf{r}_1 + \mathbf{r}_2) + \overline{M}\mathbf{s}$ is the *system center of mass*.

Thus, if $\mathbf{j}$ is a constant (or we pull Ω back to the surface $\mathbf{j} = $ constant), we get the canonical form in $(\mathbf{r}, \pi)$ and $(\mathbf{s}, \sigma)$ as pairs of conjugate variables, as expected.

Second, we compute the reduced Hamiltonian by substituting into

$$H = \frac{1}{2m}\|\mathbf{p}_1\|^2 + \frac{1}{2m}\|\mathbf{p}_2\|^2 + \frac{1}{2M}\|\mathbf{P}\|^2 + V$$

the relations

$$\mathbf{p}_1 + \mathbf{p}_2 + \mathbf{P} = \mathbf{j}, \quad \pi = \frac{1}{2}(\mathbf{p}_2 - \mathbf{p}_1) \quad \text{and} \quad \sigma = \mathbf{P} - \overline{M}\mathbf{j}.$$

One obtains

$$H = \frac{1}{4m\overline{M}}\|\sigma\|^2 + \frac{\|\pi\|^2}{m} + \frac{1}{2\mathcal{M}}\|\mathbf{j}\|^2 + V. \tag{3.2.10}$$

Notice that in this case the *reduced Hamiltonian and Lagrangian are Legendre transformations of one another.*

Remark There are other choices of canonical coordinates for the system after reduction by translations, but they are not as convenient as the above Jacobi-Bertrand-Haretu coordinates. For example, if one uses the apparently more "democratic" choice

$$\mathbf{s}_1 = \mathbf{r}_1 - \mathbf{R} \quad \text{and} \quad \mathbf{s}_2 = \mathbf{r}_2 - \mathbf{R},$$

then their conjugate momenta are

$$\pi_1 = \mathbf{p}_1 - \overline{m}\mathbf{j} \quad \text{and} \quad \pi_2 = \mathbf{p}_2 - \overline{m}\mathbf{j}.$$

One finds that the reduced Lagrangian is

$$L = \frac{1}{2}m''\|\dot{\mathbf{s}}_1\|^2 + \frac{1}{2}m''\|\dot{\mathbf{s}}_2\|^2 - m\overline{m}\dot{\mathbf{s}}_1 \cdot \dot{\mathbf{s}}_2 - V$$

and the reduced Hamiltonian is

$$H = \frac{1}{2m'}\|\pi_1\|^2 + \frac{1}{2m'}\|\pi_2\|^2 + \frac{1}{M}\pi_1 \cdot \pi_2 + V$$

where $m' = mM/(m + M)$ and $m'' = mM\overline{m}/m'$. It is the cross terms in H and L that make these expressions inconvenient. ◆

Next we turn to a main point, which is the action of the rotation group, $SO(3)$ and the role of the discrete group that "swaps" the two masses m.

Our phase space is $Z = T^*(\mathbb{R}^3 \times \mathbb{R}^3)$ parametrized by $(\mathbf{r}, \mathbf{s}, \pi, \sigma)$ with the canonical symplectic structure. The group $G = SO(3)$ acts on Z by the cotangent lift of rotations on $\mathbb{R}^3 \times \mathbb{R}^3$. The corresponding infinitesimal action on Q is

$$\xi_Q(\mathbf{r}, \mathbf{s}) = (\xi \times \mathbf{r}, \xi \times \mathbf{s})$$

and the corresponding momentum map is $\mathbf{J} : Z \to \mathbb{R}^3$, where

$$\langle \mathbf{J}(\mathbf{r}, \mathbf{s}, \pi, \sigma), \xi \rangle = \pi \cdot (\xi \times \mathbf{r}) + \sigma \cdot (\xi \times \mathbf{s}) = \xi \cdot (\mathbf{r} \times \pi + \mathbf{s} \times \sigma).$$

In terms of the original variables, we find that

$$\mathbf{J} = \mathbf{r} \times \boldsymbol{\pi} + \mathbf{s} \times \boldsymbol{\sigma} = (\mathbf{r}_2 - \mathbf{r}_1) \times \frac{1}{2}(\mathbf{p}_2 - \mathbf{p}_1) + (\mathbf{R} - \frac{1}{2}(\mathbf{r}_1 + \mathbf{r}_2)) \times (\mathbf{P} - \overline{M}\mathbf{j})$$

$$(3.2.11)$$

which becomes, after simplification,

$$\mathbf{J} = \mathbf{r}_1 \times (\mathbf{p}_1 - \overline{m}\mathbf{j}) + \mathbf{r}_2 \times (\mathbf{p}_2 - \overline{m}\mathbf{j}) + \mathbf{R} \times (\mathbf{P} - \overline{M}\mathbf{j}) \qquad (3.2.12)$$

which is the correct total angular momentum of the system.

Remark This expression for $\mathbf{J}$ also equals $\mathbf{s}_1 \times \boldsymbol{\pi}_1 + \mathbf{s}_2 \times \boldsymbol{\pi}_2$ in terms of the alternative variables discussed above. $\blacklozenge$

3.3   The Mechanical Connection

In this section we work in the general context of a simple mechanical G-system. We shall also assume that G acts freely on Q so we can regard $Q \to Q/G$ as a principal G-bundle. (We make some remarks on this assumption later.)

For each $q \in Q$, let the *locked inertia tensor* be the map $\mathbb{I}(q) : \mathfrak{g} \to \mathfrak{g}^*$ defined by,

$$\langle \mathbb{I}(q)\eta, \zeta \rangle = \langle\!\langle \eta_Q(q), \zeta_Q(q) \rangle\!\rangle. \qquad (3.3.1)$$

Since the action is free, $\mathbb{I}(q)$ is indeed an inner product. The terminology comes from the fact that for coupled rigid or elastic systems, $\mathbb{I}(q)$ is the classical moment of inertia tensor of the rigid body obtained by locking all the joints of the system. In coordinates,

$$\mathbb{I}_{ab} = g_{ij} A^i{}_a A^j{}_b. \qquad (3.3.2)$$

Define the map $\alpha : TQ \to \mathfrak{g}$ which assigns to each (q, v) the corresponding *angular velocity of the locked system*:

$$\alpha(q, v) = \mathbb{I}(q)^{-1}(\mathbf{J}(\mathbb{F}L(q, v))). \qquad (3.3.3)$$

In coordinates,

$$\alpha^a = \mathbb{I}^{ab} g_{ij} A^i{}_b v^j. \qquad (3.3.4)$$

One can think of α_q as the map induced by the momentum map via two Legendre transformations, one on the given system and one on the instantaneous rigid body associated with it, as in Figure 3.3.1

Proposition 3.1 *The map (3.3.3) is a connection on the principal G-bundle $Q \to Q/G$.*

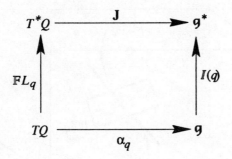

Figure 3.3.1: The diagram defining the mechanical connection.

In other words, α is G-equivariant and satisfies $\alpha(\xi_Q(q)) = \xi$, both of which are readily verified. In checking equivariance one uses invariance of the metric; *i.e.*, equivariance of $\mathbb{F}L : TQ \to T^*Q$, equivariance of $\mathbf{J} : T^*Q \to \mathfrak{g}^*$, and equivariance of $\mathbb{I}$ in the sense of a map $\mathbb{I} : Q \to L(\mathfrak{g}, \mathfrak{g}^*)$, namely $\mathbb{I}(gq) \cdot Ad_g\xi = Ad^*_{g^{-1}}\mathbb{I}(q) \cdot \xi$.

We call α the **mechanical connection**. This connection is *implicitly* used in the work of Smale [1970], and Abraham and Marsden [1978], and *explicitly* in Kummer [1981], Guichardet [1984], Shapere and Wilczek [1989], Simo, Lewis and Marsden [1991], and Montgomery [1990]. We note that all of the preceding formulas are given in Abraham and Marsden [1978, §4.5], but not from the point of view of connections.

The *horizontal space* hor_q of the connection α at $q \in Q$ is given by the kernel of α_q; thus, by (3.3.3),

$$\mathrm{hor}_q = \{(q, v) \mid \mathbf{J}(\mathbb{F}L(q, v)) = 0\}; \tag{3.3.5}$$

i.e., the space orthogonal to the G-orbits, or, equivalently, the space of states with zero total angular momentum in the case $G = SO(3)$ (see Figure 3.3.2). The vertical space consists of vectors that are mapped to zero under the projection $Q \to S = Q/G$; *i.e.*,

$$\mathrm{ver}_q = \{\xi_Q(q) \mid \xi \in \mathfrak{g}\}. \tag{3.3.6}$$

For each $\mu \in \mathfrak{g}^*$, define the 1-form α_μ on Q by

$$\langle \alpha_\mu(q), v \rangle = \langle \mu, \alpha(q, v) \rangle \tag{3.3.7}$$

i.e.,

$$(\alpha_\mu)_i = g_{ij} A^j{}_b \mu_a \mathbb{I}^{ab}. \tag{3.3.8}$$

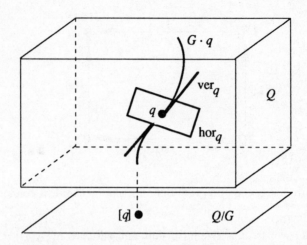

Figure 3.3.2: The horizontal space of the mechanical connection.

Proposition 3.2 α_μ *takes values in* $\mathbf{J}^{-1}(\mu)$.

Proof $\langle \mathbf{J}(\alpha_\mu(q)), \xi \rangle = \langle \alpha_\mu(q), \xi_Q(q) \rangle = \langle \mu, \alpha(\xi_Q(q)) \rangle = \langle \mu, \xi \rangle.$ ∎

Notice that this proposition holds for *any* connection on $Q \to Q/G$, not just the mechanical connection. Here we use the same notation of α_μ as in Smale [1970] and Abraham and Marsden [1978]. In these references it is proved that α_μ is characterized by

$$K(\alpha_\mu(q)) = \inf\{K(q,\beta) \mid \beta \in \mathbf{J}_q^{-1}(\mu)\} \tag{3.3.9}$$

where $\mathbf{J}_q = \mathbf{J}|T_q^*Q$ and $K(q,p) = \frac{1}{2}\|p\|_q^2$ is the kinetic energy function (see Figure 3.3.3).

The horizontal-vertical decomposition of a vector $(q,v) \in T_qQ$ is given by the general prescription

$$v = \mathrm{hor}_q v + \mathrm{ver}_q v \tag{3.3.10}$$

where

$$\mathrm{ver}_q v = [\alpha(q,v)]_Q(q) \quad \text{and} \quad \mathrm{hor}_q v = v - \mathrm{ver}_q v.$$

In terms of T^*Q rather than TQ, we define a map $\omega : T^*Q \to \mathfrak{g}$ by

$$\omega(q,p) = \mathbb{I}(q)^{-1}\mathbf{J}(q,p) \tag{3.3.11}$$

i.e.,

$$\omega^a = \mathbb{I}^{ab}A^i_{\ b}p_i, \tag{3.3.12}$$

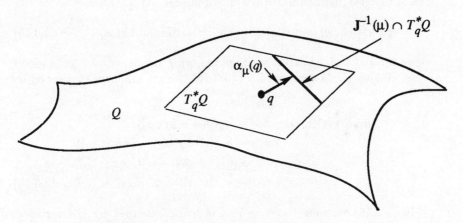

Figure 3.3.3: The mechanical connection as the orthogonal vector to the level set of **J** in the cotangent fiber.

and, using a slight abuse of notation, a projection hor : $T^*Q \to \mathbf{J}^{-1}(0)$ by

$$\text{hor}(q, p) = p - \alpha_{\mathbf{J}(q,p)}(q) \qquad (3.3.13)$$

i.e.,

$$(\text{hor}(q, p))_i = p_i - g_{ij} A^j_{\ b} p_k A^k_{\ a} \mathbb{I}^{ab}. \qquad (3.3.14)$$

This map hor in (3.3.13) will play a fundamental role in what follows. We also refer to hor as the **shifting map**. This map will be used in the proof of the cotangent bundle reduction theorem in §3.4; it is also an essential ingredient in the description of a particle in a Yang-Mills field via the Kaluza-Klein construction, generalizing the electromagnetic case in which $p \mapsto p - \frac{e}{c} A$. This aspect will be brought in later.

Example Let $Q = G$ be a Lie group with symmetry group G itself, acting say on the left. In this case, α_μ is independent of the Hamiltonian. Since $\alpha_\mu(q) \in J^{-1}(\mu)$, it follows that α_μ is the right invariant one form whose value at the identity e is μ. (This same one form α_μ was already used in Marsden and Weinstein [1974].) ◆

The **curvature** curv α of the connection α is the **covariant exterior derivative** of α, defined to be the exterior derivative acting on the horizontal components. The curvature may be regarded as a measure of the lack of integrability of the horizontal subbundle. At $q \in Q$,

$$(\text{curv}\,\alpha)(v, w) = d\alpha(\text{hor}\,v, \text{hor}\,w) = -\alpha([\text{hor}\,v, \text{hor}\,w]),\qquad(3.3.15)$$

where the Jacobi-Lie bracket is computed using extensions of v, w to vector fields. Taking the μ-component of (3.3.15), we get a 2-form $\langle \mu, \text{curv}\,\alpha\rangle$ on Q given by

$$
\begin{aligned}
\langle \mu, \text{curv}\,\alpha\rangle(v, w) &= -\alpha_\mu([\text{hor}\,v, \text{hor}\,w])\\
&= -\alpha_\mu([v - \alpha(v)_Q, w - \alpha(w)_Q])\\
&= \alpha_\mu([v, \alpha(w)_Q] - \alpha_\mu([w, \alpha(v)_Q]\\
&\quad - \alpha_\mu([v, w]) - \langle \mu, [\alpha(v), \alpha(w)]\rangle. \quad(3.3.16)
\end{aligned}
$$

In (3.3.16) we may choose v and w to be extended by G-invariance. Then $\alpha(w)_Q = \zeta_Q$ for a *fixed* ζ, so $[v, \alpha(w)_Q] = 0$, so we can replace this term by $v[\alpha_\mu(w)]$, which is also zero, noting that $v[\langle \mu, \xi\rangle] = 0$, as $\langle \mu, \xi\rangle$ is constant. Thus, (3.3.16) gives the *structure equation*:

$$\langle \mu, \text{curv}\,\alpha\rangle = d\alpha_\mu - [\alpha, \alpha]_\mu \qquad(3.3.17)$$

where $[\alpha, \alpha]_\mu(v, w) = \langle [\alpha(v), \alpha(w)]\rangle$, the bracket being the Lie algebra bracket. Formula (3.3.17) is, of course, a standard one for curvatures of principal connections.

The *amended potential* V_μ is defined by

$$V_\mu = H \circ \alpha_\mu; \qquad(3.3.18)$$

this function also plays a crucial role in what follows. In coordinates,

$$V_\mu(q) = V(q) + \frac{1}{2}\mathbb{I}^{ab}(q)\mu_a\mu_b \qquad(3.3.19)$$

or, intrinsically

$$V_\mu(q) = V(q) + \frac{1}{2}\langle \mu, \mathbb{I}(q)^{-1}\mu\rangle. \qquad(3.3.20)$$

3.4 The Geometry and Dynamics of Cotangent Bundle Reduction

Given a symplectic G-system $(P, \Omega, G, \mathbf{J}, H)$, in Chapter 2, we defined the *reduced space* P_μ by

$$P_\mu = \mathbf{J}^{-1}(\mu)/G_\mu, \qquad(3.4.1)$$

assuming μ is a regular value of $\mathbf{J}$ (or a weakly regular value) and that the G_μ action is free and proper (or an appropriate slice theorem applies) so that the quotient (3.4.1) is a manifold. The reduction theorem states that the symplectic structure Ω on P naturally induces one on P_μ; it is denoted Ω_μ.

Recall also that $P_\mu \cong P_{\mathcal{O}}$ where

$$P_{\mathcal{O}} = \mathbf{J}^{-1}(\mathcal{O})/G$$

and $\mathcal{O} \subset \mathfrak{g}^*$ is the coadjoint orbit through μ.

Next, map $\mathbf{J}^{-1}(\mathcal{O}) \to \mathbf{J}^{-1}(0)$ by the map hor given in the last section. This induces a map on the quotient spaces by equivariance; we denote it $\text{hor}_{\mathcal{O}}$:

$$\text{hor}_{\mathcal{O}} : \mathbf{J}^{-1}(\mathcal{O})/G \to \mathbf{J}^{-1}(0)/G. \tag{3.4.2}$$

Reduction at zero is simple: $\mathbf{J}^{-1}(0)/G$ is isomorphic with $T^*(Q/G)$ by the following identification: $\beta_q \in \mathbf{J}^{-1}(0)$ satisfies $\langle \beta_q, \xi_Q(q) \rangle = 0$ for all $\xi \in \mathfrak{g}$, so we can regard β_q as a one form on $T(Q/G)$.

As a set, the fiber of the map $\text{hor}_{\mathcal{O}}$ is identified with $\mathcal{O}$. Therefore, we have realized $(T^*Q)_{\mathcal{O}}$ as a coadjoint orbit bundle over $T^*(Q/G)$. The spaces are summarized in Figure 3.4.1.

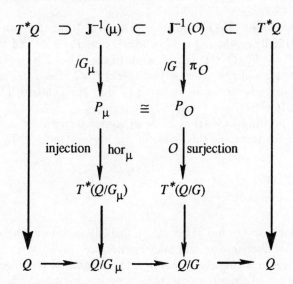

Figure 3.4.1: Cotangent bundle reduction.

The Poisson bracket structure of the bundle hor$_{\mathcal{O}}$ as a synthesis of the Lie-Poisson structure, the cotangent structure, the magnetic and interaction terms is investigated in Montgomery, Marsden and Ratiu [1984] and Montgomery [1986].

For the symplectic structure, it is a little easier to use P_μ. Here, we restrict the map hor to $\mathbf{J}^{-1}(\mu)$ and quotient by G_μ to get a map of P_μ to $\mathbf{J}^{-1}(0)/G_\mu$. If $\mathbf{J}_\mu$ denotes the momentum map for G_μ, then $\mathbf{J}^{-1}(0)/G_\mu$ embeds in $\mathbf{J}_\mu^{-1}(0)/G_\mu \cong T^*(Q/G_\mu)$. The resulting map hor$_\mu$ *embeds* P_μ into $T^*(Q/G_\mu)$. This map is induced by the shifting map:

$$p_q \mapsto p_q - \alpha_\mu(q). \tag{3.4.3}$$

The symplectic form on P_μ is obtained by restricting the form on $T^*(Q/G_\mu)$ given by

$$\Omega_{\text{canonical}} + \mathbf{d}\alpha_\mu. \tag{3.4.4}$$

The two form $\mathbf{d}\alpha_\mu$ drops to a two form β_μ on the quotient, so (3.4.4) defines the symplectic structure of P_μ. We call β_μ the *magnetic term* following the ideas from §2.10 and the terminology of Abraham and Marsden [1978]. We also note that on $\mathbf{J}^{-1}(\mu)$ (and identifying vectors and covectors via $\mathbb{F}L$), $[\alpha(v), \alpha(w)] = [\mu, \mu] = 0$, so that the form β_μ *may also be regarded as the form induced by the μ-component of the curvature.* In the $P_{\mathcal{O}}$ context, the distinction between $\mathbf{d}\alpha_\mu$ and $(\text{curv } \alpha)_\mu$ becomes important. We shall see this aspect in Chapter 4.

Two limiting cases are noteworthy. The first (that one can associate with Arnold [1966]) is when $Q = G$ in which case $P_{\mathcal{O}} \cong \mathcal{O}$ and the base is trivial in the $P_{\mathcal{O}} \to T^*(Q/G)$ picture, while in the $P_\mu \to T^*(Q/G_\mu)$ picture, the fiber is trivial and the space is just $Q/G_\mu \cong \mathcal{O}$. Here the description of the orbit symplectic structure induced by $\mathbf{d}\alpha_\mu$ coincides with that given by Kirillov [1976].

The other limiting case (that one can associate with Smale [1970]) is when $G = G_\mu$; for instance, this holds in the abelian case. Then

$$P_\mu = P_{\mathcal{O}} = T^*(Q/G)$$

with symplectic form $\Omega_{\text{canonical}} + \beta_\mu$.

Remark Montgomery has given (unpublished) an interesting condition, generalizing this abelian case and allowing non-free actions, which guarantees that P_μ is still a cotangent bundle. This condition (besides technical assumptions guaranteeing the objects in question are manifolds) is

$$\dim \mathfrak{g} - \dim \mathfrak{g}_\mu = 2(\dim \mathfrak{g}_Q - \dim \mathfrak{g}_Q^\mu) \tag{3.4.5}$$

where $\mathfrak{g}_Q$ is the isotropy algebra of the G-action on Q (assuming $\mathfrak{g}_Q$ has constant dimension) and $\mathfrak{g}_Q^\mu$ that for G_μ. The result is that P_μ is identified with $T^*(Q^\mu/G_\mu)$, where Q^μ is the projection of $\mathbf{J}^{-1}(\mu) \subset T^*Q$ to Q under the cotangent bundle projection. (For free actions of G on Q, $Q^\mu = Q$.) The proof is a modification of the one given above.

Note that the construction of P_μ requires only that the G_μ action be free; then one gets P_μ embedded in $T^*(Q/G_\mu)$. However, the bundle picture of $P_\mathcal{O} \to T^*(Q/G)$ with fiber $\mathcal{O}$ requires G to act freely on Q, so Q/G is defined. If the G-action is not free, one can either deal with singularities or use Montgomery's method above, replacing *shape space* Q/G with the modified shape space Q^μ/G (and drop (3.4.5)). For example, already in the Kepler problem in which $G = SO(3)$ acts on $Q = \mathbb{R}^3 \backslash \{0\}$, the action of G on Q is not free; for $\mu \neq 0$, Q^μ is the plane orthogonal to μ (minus $\{0\}$). On the other hand, $G_\mu = S^1$, the rotations about the axis $\mu \neq 0$, acts freely on $\mathbf{J}^{-1}(\mu)$, so P_μ is defined. But G_μ does not act freely on Q! Instead, P_μ is actually $T^*(Q^\mu/G_\mu) = T^*(0, \infty)$, with the canonical cotangent structure. The case $\mu = 0$ is singular and is more interesting! (See Marsden [1981].)
♦

Remarks Another context in which one can get interesting results, both mathematically and physically, is that of semi-direct products (Marsden, Ratiu and Weinstein [1984a,b]). Here we consider a representation of G on a vector space V. We form the semi-direct product $S = G \circledS V$ and look at its coadjoint orbit $\mathcal{O}_{\mu,a}$ through $(\mu, a) \in \mathfrak{g}^* \times V^*$. The *semi-direct product reduction theorem* says that $\mathcal{O}_{\mu,a}$ is symplectomorphic with the reduced space obtained by reducing T^*G at $\mu_a = \mu \mid \mathfrak{g}_a$ by the sub-group G_a (the isotropy for the action of G on V^* at $a \in V^*$). If G_a is abelian (the generic case) then abelian reduction gives

$$\mathcal{O}_{\mu,a} \cong T^*(G/G_a)$$

with the canonical plus magnetic structure. For example, the generic orbits in $SE(3) = SO(3) \circledS \mathbb{R}^3$ are cotangent bundles of spheres; but the orbit symplectic structure has a non-trivial magnetic term.

Semi-direct products come up in a variety of interesting physical situations, such as the heavy rigid body, compressible fluids and MHD; we refer to Marsden, Ratiu, and Weinstein [1984a,b] and Holm, Marsden, Ratiu, and Weinstein [1985] for details.

For semidirect products, one has a useful result called *reduction by stages*. Namely, one can reduce by S in two successive stages, *first* by V and *then* by G. For example, for $SE(3)$ one can reduce *first* by translations, and *then* by rotations, and the result is the same as reducing by

$SE(3)$. Coupled with the cotangent bundle reduction theorem, one finds that the generic coadjoint orbits of $SE(3)$ are cotangent bundles T^*S^2, with magnetic terms. ♦

Now we turn to the dynamics of cotangent bundle reduction. Given a symplectic G-system, we get a reduced Hamiltonian system on $P_\mu \cong P_{\mathcal{O}}$ obtained by restricting H to $\mathbf{J}^{-1}(\mu)$ or $\mathbf{J}^{-1}(\mathcal{O})$ and then passing to the quotient. This produces the reduced Hamiltonian function H_μ and thereby a Hamiltonian system on P_μ. The resulting vector field is the one obtained by restricting and projecting the Hamiltonian vector field X_H on P. The resulting dynamical system X_{H_μ} on P_μ is called the *reduced Hamiltonian system*.

Let us compute H_μ in each of the pictures P_μ and $P_{\mathcal{O}}$. In either case the shift by the map hor is basic, so let us first compute the function on $\mathbf{J}^{-1}(0)$ given by

$$H_{\alpha_\mu}(q,p) = H(q,p + \alpha_\mu(q)). \tag{3.4.6}$$

Indeed,

$$
\begin{aligned}
H_{\alpha_\mu}(q,p) &= \frac{1}{2}\langle\langle p + \alpha_\mu, p + \alpha_\mu\rangle\rangle_q + V(q) \\
&= \frac{1}{2}\|p\|_q^2 + \langle\langle p, \alpha_\mu\rangle\rangle_q + \frac{1}{2}\|\alpha_\mu\|_q^2 + V(q). \tag{3.4.7}
\end{aligned}
$$

If $p = \mathbb{F}L \cdot v$, then $\langle\langle p, \alpha_\mu\rangle\rangle_q = \langle\alpha_\mu, v\rangle = \langle\mu, \alpha(q,v)\rangle = \langle\mu, \mathbb{I}(q)\mathbf{J}(p)\rangle = 0$ since $\mathbf{J}(p) = 0$. Thus, on $\mathbf{J}^{-1}(0)$,

$$H_{\alpha_\mu}(q,p) = \frac{1}{2}\|p\|_q^2 + V_\mu(q). \tag{3.4.8}$$

In $T^*(Q/G_\mu)$, we obtain H_μ by selecting any representative (q,p) of $T^*(Q/G_\mu)$ in $\mathbf{J}^{-1}(0) \subset T^*Q$, shifting it to $\mathbf{J}^{-1}(\mu)$ by $p \mapsto p + \alpha_\mu(q)$ and then calculating H. Thus, the above calculation (3.4.8) proves:

Proposition 3.3 *The reduced Hamiltonian H_μ is the function obtained by restricting to the symplectic subbundle $P_\mu \subset T^*(Q/G_\mu)$, the function*

$$H_\mu(q,p) = \frac{1}{2}\|p\|^2 + V_\mu(q) \tag{3.4.9}$$

defined on $T^*(Q/G_\mu)$ with the symplectic structure

$$\Omega_\mu = \Omega_{\mathrm{can}} + \beta_\mu \tag{3.4.10}$$

where β_μ is the two form on Q/G_μ obtained from $d\alpha_\mu$ on Q by passing to the quotient. Here we use the quotient metric on Q/G_μ and identify V_μ with a function on Q/G_μ.

Example If $Q = G$ and the symmetry group is G itself, then $P_\mu \subset T^*(Q/G_\mu)$ sits as the zero section. In fact P_μ is identified with $Q/G_\mu \cong G/G_\mu \cong \mathcal{O}_\mu$. ♦

To describe H_μ on $\mathbf{J}^{-1}(\mathcal{O})/G$ is easy abstractly; one just calculates H restricted to $\mathbf{J}^{-1}(\mathcal{O})$ and passes to the quotient. More concretely, we choose an element $[(q,p)] \in T^*(Q/G)$, where we identify the representative with an element of $\mathbf{J}^{-1}(0)$. We also choose an element $\nu \in \mathcal{O}$, a coadjoint orbit, and shift $(q,p) \mapsto (q, p + \alpha_\nu(q))$ to a point in $\mathbf{J}^{-1}(\mathcal{O})$. Thus, we get:

Proposition 3.4 *Regarding $P_\mu \cong P_\mathcal{O}$ an an $\mathcal{O}$-bundle over $T^*(Q/G)$, the reduced Hamiltonian is given by*

$$H_\mathcal{O}(q,p,\nu) = \frac{1}{2}\|p\|^2 + V_\nu(q)$$

where (q,p) is a representative in $\mathbf{J}^{-1}(0)$ of a point in $T^*(Q/G)$ and where $\nu \in \mathcal{O}$.

The symplectic structure in this second picture was described abstractly above. To describe it concretely in terms of $T^*(Q/G)$ and $\mathcal{O}$ is more difficult; this problem was solved by Montgomery, Marsden and Ratiu [1984]. We shall see some special aspects of this structure when we separate internal and rotational modes in the next chapters.

3.5 Examples

We conclude this chapter with some basic examples of the cotangent bundle reduction construction. The examples will be the spherical pendulum, the double spherical pendulum, and the water molecule. Of course many more examples can be given and we refer the reader to the literature cited for a wealth of information. The ones we have chosen are, we hope, simple enough to be of pedagogical value, yet interesting.

Example 1 *The spherical pendulum* Here, $Q = S^2$ with the standard metric as a sphere of radius R in $\mathbb{R}^3$, V is the gravitational potential for a mass m, and $G = S^2$ acts on Q by rotations about the vertical axis. Relative to coordinates θ, φ as in Figure 3.5.1, we have $V(\theta, \varphi) = -mgR\cos\theta$.

We claim that the mechanical connection $\alpha : TQ \to \mathbb{R}$ is given by

$$\alpha(\theta, \varphi, \dot\theta, \dot\varphi) = \dot\varphi. \tag{3.5.1}$$

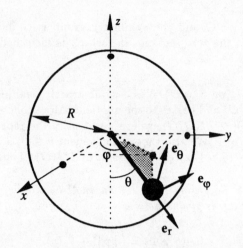

Figure 3.5.1: The configuration space of the spherical pendulum is the two sphere.

To see this, note that

$$\xi_Q(\theta, \varphi) = (\theta, \varphi, 0, \xi) \tag{3.5.2}$$

since $G = S^1$ acts by rotations about the z-axis: $(\theta, \varphi) \mapsto (\theta, \varphi + \psi)$. The metric is

$$\langle\langle(\theta, \varphi, \dot{\theta}_1, \dot{\varphi}_1), (\theta, \varphi, \dot{\theta}_2, \dot{\varphi}_2)\rangle\rangle = mR^2 \dot{\theta}_1 \dot{\theta}_2 + mR^2 \sin^2\theta \dot{\varphi}_1 \dot{\varphi}_2, \tag{3.5.3}$$

which is m times the standard inner product of the corresponding vectors in $\mathbb{R}^3$.

The momentum map is the angular momentum about the z-axis:

$$J : T^*Q \to \mathbb{R}; J(\theta, \varphi, p_\theta, p_\varphi) = p_\varphi \tag{3.5.4}$$

and the Legendre transformation is

$$p_\theta = mR^2 \dot{\theta}, \; p_\varphi = (mR^2 \sin^2\theta)\dot{\varphi} \tag{3.5.5}$$

and the locked inertia tensor is

$$\langle \mathbb{I}(\theta, \varphi)\eta, \zeta\rangle = \langle\langle(\theta, \varphi, 0, \eta)(\theta, \varphi, 0, \zeta)\rangle\rangle = (mR^2 \sin^2\theta)\eta\zeta. \tag{3.5.6}$$

Note that $\mathbb{I}(\theta, \varphi) = m(R\sin\theta)^2$ is just the (instantaneous) moment of inertia of the mass m about the z-axis.

In this example, we identify Q/S^1 with the interval $[0, \pi]$; *i.e.*, the θ-variable. In fact, it is convenient to regard Q/S^1 as S^1 mod $\mathbb{Z}_2$, where $\mathbb{Z}_2$

acts by reflection $\theta \mapsto -\theta$. This helps to understand the singularity in the quotient space.

For $\mu \in \mathfrak{g}^* \cong \mathbb{R}$, the one form α_μ is given by (3.3.3) and (3.3.7) as

$$\alpha_\mu(\theta, \varphi) = \mu d\varphi. \tag{3.5.7}$$

From (3.5.4), note that it is clear that α_μ takes values in $\mathbf{J}^{-1}(\mu)$. The shifting map is given by (3.3.13):

$$\mathrm{hor}(\theta, \varphi, p_\theta, p_\varphi) = (\theta, \varphi, p_\theta, 0).$$

The curvature of the connection α is zero in this example. The amended potential is

$$V_\mu(\theta) = V(\theta, \varphi) + \frac{1}{2}\langle \mu, \mathbb{I}(\theta, \varphi)^{-1}\mu \rangle = -mgR\cos\theta + \frac{1}{2}\frac{\mu^2}{mR^2 \sin^2\theta}$$

and so the reduced Hamiltonian on $T^*(S^1/\mathbb{Z}_2)$ is

$$H_\mu(\theta, p_\theta) = \frac{1}{2}\frac{p_\theta^2}{mR^2} + V_\mu(\theta). \tag{3.5.8}$$

The reduced Hamiltonian equations are therefore

$$\dot{\theta} = \frac{p_\theta}{mR^2},$$

and

$$\dot{p} = mgR\sin\theta - \frac{\cos\theta}{\sin^2\theta}\frac{\mu^2}{mR^2}. \tag{3.5.9}$$

Note that the extra term is singular at $\theta = 0$. ◆

Example 2 *The double spherical pendulum*. Consider the mechanical system consisting of two coupled spherical pendula in a gravitational field (see Figure 3.5.2).

Let the position vectors of the individual pendula relative to their joints be denoted $\mathbf{q}_1$ and $\mathbf{q}_2$ with fixed lengths l_1 and l_2 and with masses m_1 and m_2. The configuration space is $Q = S_{l_1}^2 \times S_{l_2}^2$, the product of spheres of radii l_1 and l_2 respectively. Since the "absolute" vector for the second pendulum is $\mathbf{q}_1 + \mathbf{q}_2$, the Lagrangian is

$$
\begin{aligned}
L(\mathbf{q}_1, \mathbf{q}_2, \dot{\mathbf{q}}_1, \dot{\mathbf{q}}_2) = {} & \frac{1}{2}m_1\|\dot{\mathbf{q}}_1\|^2 + \frac{1}{2}m_2\|\dot{\mathbf{q}}_1 + \dot{\mathbf{q}}_2\|^2 \\
& - m_1 g \mathbf{q}_1 \cdot \mathbf{k} - m_2 g(\mathbf{q}_1 + \mathbf{q}_2) \cdot \mathbf{k}. \tag{3.5.10}
\end{aligned}
$$

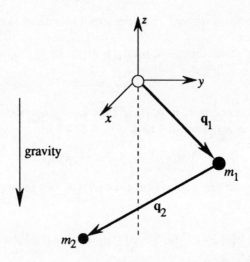

Figure 3.5.2: The configuration space for the double spherical pendulum consists of two copies of the two sphere.

Note that (3.5.10) has the standard form of kinetic minus potential energy. We identify the velocity vectors $\dot{\mathbf{q}}_1$ and $\dot{\mathbf{q}}_2$ with vectors perpendicular to $\mathbf{q}_1$ and $\mathbf{q}_2$, respectively.

The conjugate momenta are

$$\mathbf{p}_1 = \frac{\partial L}{\partial \dot{\mathbf{q}}_1} = m_1\dot{\mathbf{q}}_1 + m_2(\dot{\mathbf{q}}_1 + \dot{\mathbf{q}}_2) \qquad (3.5.11)$$

and

$$\mathbf{p}_2 = \frac{\partial L}{\partial \dot{\mathbf{q}}_2} = m_2(\dot{\mathbf{q}}_1 + \dot{\mathbf{q}}_2) \qquad (3.5.12)$$

regarded as vectors in $\mathbb{R}^3$ that are only paired with vectors orthogonal to $\mathbf{q}_1$ and $\mathbf{q}_2$ respectively.

The Hamiltonian is therefore

$$H(\mathbf{q}_1, \mathbf{q}_2, \mathbf{p}_1, \mathbf{p}_2) = \frac{1}{2m_1}\|\mathbf{p}_1 - \mathbf{p}_2\|^2 + \frac{1}{2m_2}\|\mathbf{p}_2\|^2$$
$$+ m_1 g\mathbf{q}_1 \cdot \mathbf{k} + m_2 g(\mathbf{q}_1 + \mathbf{q}_2) \cdot \mathbf{k}. \quad (3.5.13)$$

The equations of motion are the Euler-Lagrange equations for L or, equivalently, Hamilton's equations for H. To write them out explicitly, it is easiest to coordinatize Q. We will describe this process in §5.5.

Now let $G = S^1$ act on Q by simultaneous rotation about the z-axis. If R_θ is the rotation by an angle θ, the action is

$$(\mathbf{q}_1, \mathbf{q}_2) \mapsto (R_\theta \mathbf{q}_1, R_\theta \mathbf{q}_2).$$

The infinitesimal generator corresponding to the rotation vector $\omega \mathbf{k}$ is $\omega(\mathbf{k} \times \mathbf{q}_1, \mathbf{k} \times \mathbf{q}_2)$ and so the momentum map is

$$
\begin{aligned}
\langle \mathbf{J}(\mathbf{q}_1, \mathbf{q}_2, \mathbf{p}_1, \mathbf{p}_2), \omega \mathbf{k} \rangle &= \omega[\mathbf{p}_1 \cdot (\mathbf{k} \times \mathbf{q}_1) + \mathbf{p}_2 \cdot (\mathbf{k} \times \mathbf{q}_2)] \\
&= \omega \mathbf{k} \cdot [\mathbf{q}_1 \times \mathbf{p}_1 + \mathbf{q}_2 \times \mathbf{p}_2]
\end{aligned}
$$

i.e.,

$$\mathbf{J} = \mathbf{k} \cdot [\mathbf{q}_1 \times \mathbf{p}_1 + \mathbf{q}_2 \times \mathbf{p}_2]. \tag{3.5.14}$$

Note that from (3.5.11) and (3.5.12),

$$
\begin{aligned}
\mathbf{J} &= \mathbf{k} \cdot [m_1 \mathbf{q}_1 \times \dot{\mathbf{q}}_1 + m_2 \mathbf{q}_1 \times (\dot{\mathbf{q}}_1 + \dot{\mathbf{q}}_2) + m_2 \mathbf{q}_2 \times (\dot{\mathbf{q}}_1 + \dot{\mathbf{q}}_2)] \\
&= \mathbf{k} \cdot [m_1 (\mathbf{q}_1 \times \dot{\mathbf{q}}_1) + m_2 (\mathbf{q}_1 + \mathbf{q}_2) \times (\dot{\mathbf{q}}_1 + \dot{\mathbf{q}}_2)].
\end{aligned}
$$

The locked inertia tensor is read off from the metric defining L in (3.5.10):

$$
\begin{aligned}
\langle \mathbb{I}(\mathbf{q}_1, \mathbf{q}_2) \omega_1 \mathbf{k}, \omega_2 \mathbf{k} \rangle &= \omega_1 \omega_2 \langle ((\mathbf{k} \times \mathbf{q}_1, \mathbf{k} \times \mathbf{q}_2), (\mathbf{k} \times \mathbf{q}_1, \mathbf{k} \times \mathbf{q}_2) \rangle \\
&= \omega_1 \omega_2 \{ m_1 \|\mathbf{k} \times \mathbf{q}_1\|^2 + m_2 \|\mathbf{k} \times (\mathbf{q}_1 + \mathbf{q}_2)\|^2 \}.
\end{aligned}
$$

Using the identity $(\mathbf{k} \times \mathbf{r}) \cdot (\mathbf{k} \times \mathbf{s}) = \mathbf{r} \cdot \mathbf{s} - (\mathbf{k} \cdot \mathbf{r})(\mathbf{k} \cdot \mathbf{s})$, we get

$$\mathbb{I}(\mathbf{q}_1, \mathbf{q}_2) = m_1 \|\mathbf{q}_1^\perp\|^2 + m_2 \|(\mathbf{q}_1 + \mathbf{q}_2)^\perp\|^2 \tag{3.5.15}$$

where $\|\mathbf{q}_1^\perp\|^2 = \|\mathbf{q}_1\|^2 - \|\mathbf{q}_1 \cdot \mathbf{k}\|^2$ is the square length of the projection of $\mathbf{q}_1$ onto the xy-plane. Note that $\mathbb{I}$ is just the moment of inertia of the system about the $\mathbf{k}$-axis.

The mechanical connection is given by (3.3.3):

$$
\begin{aligned}
\alpha(\mathbf{q}_1, \mathbf{q}_2, \mathbf{v}_1, \mathbf{v}_2) &= \mathbb{I}^{-1} \mathbf{J}(m_1 \mathbf{v}_1 + m_2 (\mathbf{v}_1 + \mathbf{v}_2), m_2 (\mathbf{v}_1 + \mathbf{v}_2)) \\
&= \mathbb{I}^{-1} (\mathbf{k} \cdot [m_1 \mathbf{q}_1 \times \mathbf{v}_1 + m_2 (\mathbf{q}_1 + \mathbf{q}_2) \times (\mathbf{v}_1 + \mathbf{v}_2)]) \\
&= \frac{\mathbf{k} \cdot [m_1 \mathbf{q}_1 \times \mathbf{v}_1 + m_2 (\mathbf{q}_1 + \mathbf{q}_2) \times (\mathbf{v}_1 + \mathbf{v}_2)]}{m_1 \|\mathbf{q}_1^\perp\|^2 + m_2 \|(\mathbf{q}_1 + \mathbf{q}_2)^\perp\|^2}.
\end{aligned}
$$

As a one form,

$$
\begin{aligned}
\alpha(\mathbf{q}_1, \mathbf{q}_2) &= \frac{1}{m_1 \|\mathbf{q}_1^\perp\|^2 + m_2 \|(\mathbf{q}_1 + \mathbf{q}_2)^\perp\|^2} \times \\
&\quad [\mathbf{k} \times ((m_1 + m_2) \mathbf{q}_1^\perp + m_2 \mathbf{q}_2^\perp), \mathbf{k} \times m_2 (\mathbf{q}_1 + \mathbf{q}_2)^\perp].
\end{aligned}
$$

$$\tag{3.5.16}$$

A short calculation shows that $\mathbf{J}(\alpha(\mathbf{q}_1, \mathbf{q}_2)) = 1$, as it should. Also, from (3.5.16), and (3.3.20),

$$
\begin{aligned}
V_\mu(\mathbf{q}_1, \mathbf{q}_2) \;=\;& m_1 g \mathbf{q}_1 \cdot \mathbf{k} + m_2 g(\mathbf{q}_1 + \mathbf{q}_2) \cdot \mathbf{k} \\
&+ \frac{1}{2} \frac{\mu^2}{m_1 \|\mathbf{q}_1^\perp\|^2 + m_2 \|(\mathbf{q}_1 + \mathbf{q}_2)^\perp\|^2}.
\end{aligned}
\tag{3.5.17}
$$

The reduced space is $T^*(Q/S^1)$, which is 6 dimensional and it carries a nontrivial magnetic term obtained by taking the differential of (3.5.16).
◆

Example 3 *The water (or ozone) molecule.* We now compute the locked inertia tensor, the mechanical connection and the amended potential for this example at symmetric states. Since $\mathbb{I}(q) : \mathfrak{g} \to \mathfrak{g}^*$ satisfies

$$
\langle \mathbb{I}(q)\eta, \xi \rangle = \langle\langle \eta_Q(q), \xi_Q(q) \rangle\rangle
$$

where $\langle\langle\,,\,\rangle\rangle$ is the kinetic energy metric coming from the Lagrangian, $\mathbb{I}(\mathbf{r}, \mathbf{s})$ is the 3×3 matrix determined by

$$
\langle \mathbb{I}(\mathbf{r}, \mathbf{s})\eta, \xi \rangle = \langle\langle (\eta \times \mathbf{r}, \eta \times \mathbf{s}), (\xi \times \mathbf{r}, \xi \times \mathbf{s}) \rangle\rangle.
$$

From our expression for the Lagrangian, we have

$$
\langle\langle (\mathbf{u}_1, \mathbf{u}_2), (\mathbf{v}_1, \mathbf{v}_2) \rangle\rangle = \frac{m}{2}(\mathbf{u}_1 \cdot \mathbf{v}_1) + 2M\overline{m}(\mathbf{u}_2 \cdot \mathbf{v}_2)
$$

Thus,

$$
\begin{aligned}
\langle \mathbb{I}(\mathbf{r}, \mathbf{s})\eta, \xi \rangle \;=\;& \langle\langle (\eta \times \mathbf{r}, \eta \times \mathbf{s}), (\xi \times \mathbf{r}, \xi \times \mathbf{s}) \rangle\rangle \\
=\;& \frac{m}{2}(\eta \times \mathbf{r}) \cdot (\xi \times \mathbf{r}) + 2M\overline{m}(\eta \times \mathbf{s}) \cdot (\xi \times \mathbf{s}) \\
=\;& \frac{m}{2}\xi \cdot [\mathbf{r} \times (\eta \times \mathbf{r})] + 2M\overline{m}\xi \cdot [\mathbf{s} \times (\eta \times \mathbf{s})]
\end{aligned}
$$

and so

$$
\begin{aligned}
\mathbb{I}(\mathbf{r}, \mathbf{s})\eta \;=\;& \frac{m}{2}\mathbf{r} \times (\eta \times \mathbf{r}) + 2M\overline{m}\mathbf{s} \times (\eta \times \mathbf{s}) \\
=\;& \frac{m}{2}[\eta\|\mathbf{r}\|^2 - \mathbf{r}(\mathbf{r} \cdot \eta)] + 2M\overline{m}[\eta\|\mathbf{s}\|^2 - \mathbf{s}(\mathbf{s} \cdot \eta)] \\
=\;& \left[\frac{m}{2}\|\mathbf{r}\|^2 + 2M\overline{m}\|\mathbf{s}\|^2 \right] \eta - \left[\frac{m}{2}\mathbf{r}(\mathbf{r} \cdot \eta) + 2M\overline{m}\mathbf{s}(\mathbf{s} \cdot \eta) \right].
\end{aligned}
$$

Therefore,

$$
\mathbb{I}(\mathbf{r}, \mathbf{s}) = \left[\frac{m}{2}\|\mathbf{r}\|^2 + 2M\overline{m}\|\mathbf{s}\|^2 \right] Id - \left[\frac{m}{2}\mathbf{r} \otimes \mathbf{r} + 2M\overline{m}\mathbf{s} \otimes \mathbf{s} \right],
\tag{3.5.18}
$$

where Id is the identity. To form the mechanical connection we need to invert $\mathbb{I}(\mathbf{r}, \mathbf{s})$. To do so, one has to solve

$$\left[\frac{m}{2}\|\mathbf{r}\|^2 + 2M\overline{m}\|\mathbf{s}\|^2\right]\eta - \left[\frac{m}{2}\mathbf{r}(\mathbf{r} \cdot \eta) + 2M\overline{m}\mathbf{s}(\mathbf{s} \cdot \eta)\right] = \mathbf{u} \qquad (3.5.19)$$

for η.

An especially easy case (that is perhaps also the most interesting), is the case of symmetric molecules when $\mathbf{r}$ and $\mathbf{s}$ are orthogonal. Write η in the corresponding orthonormal basis as:

$$\eta = a\mathbf{r} + b\mathbf{s} + c(\mathbf{r} \times \mathbf{s}). \qquad (3.5.20)$$

Taking the dot product of (3.5.19) with $\mathbf{r}, \mathbf{s}$ and $\mathbf{r} \times \mathbf{s}$ respectively, we find

$$a = \frac{\mathbf{u} \cdot \mathbf{r}}{2M\overline{m}\|\mathbf{r}\|^2\|\mathbf{s}\|^2}, \quad b = \frac{2\mathbf{u} \cdot \mathbf{s}}{m\|\mathbf{r}\|^2\|\mathbf{s}\|^2}, \quad c = \frac{\mathbf{u} \cdot (\mathbf{r} \times \mathbf{s})}{\gamma\|\mathbf{r}\|\|\mathbf{s}\|} \qquad (3.5.21)$$

where $\gamma = \gamma(\mathbf{r}, \mathbf{s}) = \frac{m}{2}\|\mathbf{r}\|^2 + 2M\overline{m}\|\mathbf{s}\|^2$. With $r = \|\mathbf{r}\|$ and $s = \|\mathbf{s}\|$, we arrive at

$$\mathbb{I}(\mathbf{r}, \mathbf{s})^{-1} = \frac{1}{\gamma}Id + \frac{m}{4M\overline{m}r^2s^2\gamma}\mathbf{r} \otimes \mathbf{r} + \frac{4M\overline{m}}{mr^2s^2\gamma}\mathbf{s} \otimes \mathbf{s}. \qquad (3.5.22)$$

We state the general case. Let $\mathbf{s}^\perp$ be the projection of $\mathbf{s}$ along the line perpendicular to $\mathbf{r}$, so $\mathbf{s} = \mathbf{s}^\perp + \dfrac{\mathbf{s} \cdot \mathbf{r}}{r^2}\mathbf{r}$. Then

$$\mathbb{I}(\mathbf{r}, \mathbf{s})^{-1}\mathbf{u} = \eta, \qquad (3.5.23)$$

where

$$\eta = a\mathbf{r} + b\mathbf{s}^\perp + c\mathbf{r} \times \mathbf{s}^\perp$$

and

$$a = \frac{\delta\mathbf{u} \cdot \mathbf{r} + \beta\mathbf{u} \cdot \mathbf{s}^\perp}{\alpha\delta - \beta^2}, \quad b = \frac{\beta\mathbf{u} \cdot \mathbf{r} + \alpha\mathbf{u} \cdot \mathbf{s}^\perp}{\alpha\delta - \beta^2}, \quad c = \frac{\mathbf{u} \cdot (\mathbf{r} \times \mathbf{s}^\perp)}{\gamma r s^\perp}$$

and

$$\alpha = \gamma r^2 - \frac{1}{2}mr^4 - 2M\overline{m}(\mathbf{s} \cdot \mathbf{r})^2, \beta = 2M\overline{m}(\mathbf{s}^\perp)^2\mathbf{s} \cdot \mathbf{r}, \delta = \gamma r^2 - 2M\overline{m}(\mathbf{s}^\perp)^4.$$

The mechanical connection, defined in general by

$$\alpha : TQ \to \mathfrak{g}; \quad \alpha(v_q) = \mathbb{I}(q)^{-1}\mathbf{J}(\mathbb{F}L(v_q))$$

gives the angular velocity of the system. For us, let $(\mathbf{r}, \mathbf{s}, \dot{\mathbf{r}}, \dot{\mathbf{s}})$ be a tangent vector and $(\mathbf{r}, \mathbf{s}, \pi, \sigma)$ the corresponding momenta given by the Legendre transform. Then, α is determined by

$$\mathbb{I} \cdot \alpha = \mathbf{r} \times \pi + \mathbf{s} \times \sigma$$

or,

$$\left[\frac{m}{2}\|\mathbf{r}\|^2 + 2M\overline{m}\|\mathbf{s}\|^2\right]\alpha - \left[\frac{m}{2}\mathbf{r}(\mathbf{r} \cdot \alpha) + 2M\overline{m}\mathbf{s}(\mathbf{s} \cdot \alpha)\right] = \mathbf{r} \times \pi + \mathbf{s} \times \sigma.$$

The formula for α is especially simple for symmetric molecules, when $\mathbf{r} \perp \mathbf{s}$. Then one finds:

$$\alpha = \frac{\dot{\mathbf{r}} \cdot \mathbf{n}}{\|\mathbf{s}\|}\hat{\mathbf{r}} + \frac{\dot{\mathbf{s}} \cdot \mathbf{n}}{\|\mathbf{r}\|}\hat{\mathbf{s}} + \frac{1}{\gamma}\left\{\frac{m}{2}\|\mathbf{r}\|^2\mathbf{s} \cdot \dot{\mathbf{r}} - 2M\overline{m}\|\mathbf{s}\|^2\mathbf{r} \cdot \dot{\mathbf{s}}\right\}\mathbf{n} \qquad (3.5.24)$$

where $\hat{\mathbf{r}} = \mathbf{r}/\|\mathbf{r}\|, \hat{\mathbf{s}} = \mathbf{s}/\|\mathbf{s}\|$ and $\mathbf{n} = (\mathbf{r} \times \mathbf{s})/\|\mathbf{r}\|\|\mathbf{s}\|$. The formula for a general $\mathbf{r}$ and $\mathbf{s}$ is a bit more complicated, requiring the general formula (3.5.23) for $\mathbb{I}(\mathbf{r}, \mathbf{s})^{-1}$.

The connection α is defined by declaring the horizontal spaces to be $\langle\langle , \rangle\rangle$ orthogonal to the $SO(3)$-orbits; *i.e.*, spaces with zero angular momentum.

Fix a total angular momentum vector μ and let $\langle\alpha_\mu(q), v_q\rangle = \langle\mu, \alpha(v_q)\rangle$, so $\alpha_\mu(q) \in T^*Q$. For example, using α above (for symmetric molecules), we get

$$\alpha_\mu = \left[\frac{\hat{\mathbf{r}} \cdot \mu}{\|\mathbf{s}\|} + \frac{\hat{\mathbf{s}} \cdot \mu}{\|\mathbf{r}\|}\right]\mathbf{n} + \frac{m}{2\gamma}(\mathbf{n} \cdot \mu)\|\mathbf{r}\|^2\|\mathbf{s}\|\hat{\mathbf{s}} - \frac{1}{\gamma}(\mathbf{n} \cdot \mu)\|\mathbf{s}\|^2\|\mathbf{r}\| \cdot 2M\overline{m}\hat{\mathbf{s}}.$$

The momentum α_μ has the property that $\mathbf{J}(\alpha_\mu) = \mu$, as may be checked. The amended potential is then

$$V_\mu(\mathbf{r}, \mathbf{s}) = H(\alpha_\mu(\mathbf{r}, \mathbf{s})).$$

Notice that for the water molecule, the reduced bundle $(T^*Q)_\mu \to T^*S$ is an S^2 bundle over shape space. The sphere represents the effective "body angular momentum variable".   ◆

3.6   Lagrangian Reduction

So far we have concentrated on the theory of reduction for Hamiltonian systems. There is a similar procedure for Lagrangian systems, although it is not so well known. The abelian version of this was known to Routh by around 1860 and this is reported in the book of Arnold [1988]. However, that procedure has some difficulties — it is coordinate dependent, and does not apply when the magnetic term is not exact. The procedure developed here (from Marsden and Scheurle [1992]) avoids these difficulties and extends the method to the nonabelian case. We do this by including conservative gyroscopic forces into the variational principle in the sense of Lagrange and d'Alembert. One uses a Dirac constraint construction to include the cases in which the reduced space is not a tangent bundle (but it is a Dirac constraint set inside one). Some of the ideas of this section are already found in Cendra, Ibort and Marsden [1987]. The nonabelian case is well illustrated by the rigid body.

Given $\mu \in \mathfrak{g}^*$, define the *Routhian* $R^\mu : TQ \to \mathbb{R}$ by:

$$R^\mu(q, v) = L(q, v) - \langle \alpha(q, v), \mu \rangle \tag{3.6.1}$$

where α is the mechanical connection. Notice that the Routhian has the form of a *Lagrangian with a gyroscopic term*; see Bloch, Krishnaprasad, Marsden and Sánchez [1991] and Krishnaprasad and Wang [1991] for information on the use of gyroscopic systems in control theory. One can regard R^μ as a partial Legendre transform of L, changing the angular velocity $\alpha(q, v)$ to the angular momentum μ.

A basic observation about the Routhian is that solutions of the Euler-Lagrange equations for L can be regarded as solutions of the Euler-Lagrange equations for the Routhian, with the addition of "magnetic forces". To understand this statement, recall that we define the *magnetic two form* β to be

$$\beta = \mathbf{d}\alpha_\mu, \tag{3.6.2}$$

a two form on Q (that drops to Q/G_μ). In coordinates,

$$\beta_{ij} = \frac{\partial \alpha_j}{\partial q^i} - \frac{\partial \alpha_i}{\partial q^j},$$

where we write $\alpha_\mu = \alpha_i dq^i$ and

$$\beta = \sum_{i<j} \beta_{ij} \, dq^i \wedge dq^j. \tag{3.6.3}$$

We say that $q(t)$ *satisfies the Euler-Lagrange equations for a Lagrangian* $\mathcal{L}$ *with the magnetic term* β provided that the associated variational principle in the sense of Lagrange and d'Alembert is satisfied:

$$\delta \int_a^b \mathcal{L}(q(t), \dot{q}) dt = \int_a^b \mathbf{i}_{\dot{q}} \beta, \tag{3.6.4}$$

where the variations are over curves in Q with fixed endpoints and where $\mathbf{i}_{\dot{q}}$ denotes the interior product by $\dot{q}$. This condition is equivalent to the coordinate condition stating that the *Euler-Lagrange equations with gyroscopic forcing* are satisfied:

$$\frac{d}{dt}\frac{\partial \mathcal{L}}{\partial \dot{q}^i} - \frac{\partial \mathcal{L}}{\partial q^i} = \dot{q}^j \beta_{ij}. \tag{3.6.5}$$

Proposition 3.5 *A curve $q(t)$ in Q whose tangent has momentum $\mathbf{J}(q, \dot{q}) = \mu$ is a solution of the Euler-Lagrange equations for the Lagrangian L iff it is a solution of the Euler-Lagrange equations for the Routhian R^μ with gyroscopic forcing given by β.*

Proof Let p denote the momentum conjugate to q for the Lagrangian L (so in coordinates, $p_i = g_{ij}\dot{q}^j$) and let $\mathfrak{p}$ be the corresponding conjugate momentum for the Routhian. Clearly, p and $\mathfrak{p}$ are related by the momentum shift $\mathfrak{p} = p - \alpha_\mu$. Thus by the chain rule, $\frac{d}{dt}\mathfrak{p} = \frac{d}{dt}p - T\alpha_\mu \cdot \dot{q}$, or in coordinates,

$$\frac{d}{dt}p_i = \frac{d}{dt}\mathfrak{p}_i - \frac{\partial \alpha_i}{\partial q^j}\dot{q}^j. \tag{3.6.6}$$

Likewise, $\mathbf{D}_q R^\mu = \mathbf{D}_q L - \mathbf{D}_q \langle \alpha(q, v), \mu \rangle$ or in coordinates,

$$\frac{\partial R^\mu}{\partial q^i} = \frac{\partial L}{\partial q^i} - \frac{\partial \alpha_j}{\partial q^i}\dot{q}^j \tag{3.6.7}$$

Subtracting these expressions, one finds (in coordinates, for convenience):

$$\begin{aligned}
\frac{d}{dt}\frac{\partial R^\mu}{\partial \dot{q}^i} - \frac{\partial R^\mu}{\partial q^i} &= \frac{d}{dt}\frac{\partial L}{\partial \dot{q}^i} - \frac{\partial L}{\partial q^i} + \left(\frac{\partial \alpha_j}{\partial q^i} - \frac{\partial \alpha_i}{\partial q^j}\right)\dot{q}^j \\
&= \frac{d}{dt}\frac{\partial L}{\partial \dot{q}^i} - \frac{\partial L}{\partial q^i} + \beta_{ij}\dot{q}^j,
\end{aligned} \tag{3.6.8}$$

which proves the result. ∎.

Proposition 3.6 *For all $(q, v) \in TQ$ and $\mu \in \mathfrak{g}^*$ we have*

$$R^\mu = \frac{1}{2}\|\mathrm{hor}(q, v)\|^2 + \langle \mathbf{J}(q, v) - \mu, \xi \rangle - \left(V + \frac{1}{2}\langle \mathbb{I}(q)\xi, \xi \rangle\right) \tag{3.6.9}$$

where $\xi = \alpha(q, v)$.

Proof Use the definition $\mathrm{hor} = v - \xi_Q(q, v)$ and expand the square using the definition of $\mathbf{J}$. ∎

Before describing the actual reduction procedure, we relate our Routhian with the classical one. If one has an abelian group G and can identify the symmetry group by a set of cyclic coordinates, then there is a simple formula which relates R^μ to the "classical" Routhian R^μ_{class}. In this case, we assume that G is the torus T^k (or a torus cross Euclidean space) and acts on Q by $q^\alpha \mapsto q^\alpha (\alpha = 1, \cdots, m)$ and $\theta^a \mapsto \theta^a + \varphi^a (a = 1, \cdots, k)$ with $\varphi^a \in [0, 2\pi)$, where $q^1, \cdots, q^m, \theta^1, \cdots, \theta^k$ are suitably chosen (local) coordinates on Q. Then G-invariance implies that the Lagrangian $L = L(q, \dot{q}, \theta)$ does not explicitly depend on the variables θ^a, *i.e.*, these variables are *cyclic*. Moreover, the infinitesimal generator ξ_Q of $\xi = (\xi^1, \cdots, \xi^k) \in \mathfrak{g}$ on Q is given by $\xi_Q = (0, \cdots, 0, \xi^1, \cdots, \xi^k)$, and the momentum map $\mathbf{J}$ has components given by $J_a = \partial L / \partial \dot{\theta}^a$, *i.e.*,

$$J_a(q, \dot{q}, \dot{\theta}) = g_{\alpha a}(q)\dot{q}^\alpha + g_{ba}(q)\dot{\theta}^b. \tag{3.6.10}$$

Thus, given $\mu \in \mathfrak{g}^*$, the **classical Routhian** is defined by (see, for example, Arnold [1988]).

$$R^\mu_{\text{class}}(q, \dot{q}) = [L(q, \dot{q}, \dot{\theta}) - \mu_a \dot{\theta}^a]_{|\dot{\theta}^a = \dot{\theta}^a(q, \dot{q})}, \tag{3.6.11}$$

where

$$\dot{\theta}^a(q, \dot{q}) = [\mu_c - g_{\alpha c}(q)\dot{q}^\alpha]\mathbb{I}^{ca}(q) \tag{3.6.12}$$

is the unique solution of $J_a(q, \dot{q}, \dot{\theta}) = \mu_a$ with respect to $\dot{\theta}^a$.

Proposition 3.7 $R^\mu_{\text{class}} = R^\mu + \mu_c g_{\alpha a} \dot{q}^\alpha \mathbb{I}^{ca}$

Proof In the present coordinates we have

$$L = \frac{1}{2} g_{\alpha\beta}(q)\dot{q}^\alpha \dot{q}^\beta + g_{\alpha a}(q)\dot{q}^\alpha \dot{\theta}^a + \frac{1}{2} g_{ab}(q)\dot{\theta}^a \dot{\theta}^b - V(q, \theta) \tag{3.6.13}$$

and

$$\alpha_\mu = \mu_a d\theta^a + g_{b\alpha}\mu_a \mathbb{I}^{ab} dq^\alpha. \tag{3.6.14}$$

By (3.3.14), and the preceding equation, we get

$$\|\text{hor}_{(q,\theta)}(\dot{q}, \dot{\theta})\|^2 = g_{\alpha\beta}(q)\dot{q}^\alpha \dot{q}^\beta - g_{\alpha a}(q)g_{b\gamma}(q)\dot{q}^\alpha \dot{q}^\gamma \mathbb{I}^{ab}(q). \tag{3.6.15}$$

Using this and the identity $(\mathbb{I}^{ab}) = (g_{ab})^{-1}$, the proposition follows from the above definitions of R^μ and R^μ_{class} by a straightforward algebraic computation. ∎

Now we are ready to drop the variational principle (3.6.4) to the quotient space Q/G_μ, with $\mathcal{L} = R^\mu$. In this principle, the variation of the integral of R^μ is taken over curves satisfying the fixed endpoint condition; this variational principle therefore holds in particular if the curves are also constrained to satisfy the condition $\mathbf{J}(q,v) = \mu$. Then we find that the variation of the function R^μ restricted to the level set of $\mathbf{J}$ satisfies the variational condition. The restriction of R^μ to the level set equals

$$R^\mu = \frac{1}{2}\|\mathrm{hor}(q,v)\|^2 - V_\mu \qquad (3.6.16)$$

In this constrained variational principle, the endpoint conditions can be relaxed to the condition that the ends lie on *orbits* rather than be fixed. This is because the kinetic part now just involves the horizontal part of the velocity, and so the endpoint conditions in the variational principle, which involve the contraction of the momentum $\mathbf{p}$ with the variation of the configuration variable δq, vanish if $\delta q = \zeta_Q(q)$ for some $\zeta \in \mathfrak{g}$, *i.e.*, if the variation is tangent to the orbit. The condition that (q,v) be in the μ level set of $\mathbf{J}$ means that the momentum $\mathbf{p}$ vanishes when contracted with an infinitesimal generator on Q.

We note, for correlation with Chapter 7 that the term $\frac{1}{2}\|\mathrm{hor}(q,v)\|^2$ is called the Wong kinetic term and that it is closely related to the Kaluza-Klein construction.

From (3.6.10), we see that the function R^μ restricted to the level set defines a function on the quotient space $T(Q/G_\mu)$ — that is, it factors through the tangent of the projection map $\tau_\mu : Q \to Q/G_\mu$. The variational principle also drops, therefore, since the curves that join orbits correspond to those that have fixed endpoints on the base. Note, also, that the magnetic term defines a well-defined two form on the quotient as well, as is known from the Hamiltonian case, even though α_μ does not drop to the quotient in general. We have proved:

Proposition 3.8 *If $q(t)$ satisfies the Euler-Lagrange equations for L with $\mathbf{J}(q,\dot{q}) = \mu$, then the induced curve on Q/G_μ satisfies the reduced Lagrangian variational principle; that is, the variational principle of Lagrange and d'Alembert on Q/G_μ with magnetic term β and the Routhian dropped to $T(Q/G_\mu)$.*

In the special case of a torus action, i.e., with cyclic variables, this reduced variational principle is equivalent to the Euler-Lagrange equations for the classical Routhian, which agrees with the classical procedure of Routh.

For the rigid body, we get a variational principle for curves on the momentum sphere — here $Q/G_\mu \cong S^2$. For it to be well defined, it is essential that one uses the variational principle in the sense of Lagrange and d'Alembert, and not in the naive sense of the Lagrange-Hamilton principle.

In this case, one checks that the dropped Routhian is just (up to a sign) the kinetic energy of the body in body coordinates. The principle then says that the variation of the kinetic energy over curves with fixed points on the two sphere equals the integral of the magnetic term (in this case a factor times the area element) contracted with the tangent to the curve. One can check that this is correct by a direct verification. (If one wants a variational principle in the usual sense, then one can do this by introduction of "Clebsch variables", as in Marsden and Weinstein [1983] and Cendra and Marsden [1987].)

The rigid body also shows that the reduced variational principle given by Proposition 3.8 in general is degenerate. This can be seen in two essentially equivalent ways; first, the projection of the constraint $\mathbf{J} = \mu$ can produce a nontrivial condition in $T(Q/G_\mu)$ — corresponding to the embedding as a symplectic subbundle of P_μ in $T^*(Q/G_\mu)$. For the case of the rigid body, the subbundle is the zero section, and the symplectic form is all magnetic (*i.e.*, all coadjoint orbit structure). The second way to view it is that the kinetic part of the induced Lagrangian is degenerate in the sense of Dirac, and so one has to cut it down to a smaller space to get well defined dynamics. In this case, one cuts down the metric corresponding to its degeneracy, and this is, coincidentally, the same cutting down as one gets by imposing the constraint coming from the image of $\mathbf{J} = \mu$ in the set $T(Q/G_\mu)$.

For the rigid body, and more generally, for T^*G, the one form α_μ is independent of the Lagrangian, or Hamiltonian. It is in fact, the right invariant one form on G equaling μ at the identity, the same form used by Marsden and Weinstein [1974] in the identification of the reduced space. Moreover, the system obtained by the Lagrangian reduction procedure above is "already Hamiltonian"; in this case, the reduced symplectic structure is "all magnetic".

There is a well defined *reconstruction procedure* for these systems. One can horizontally lift a curve in Q/G_μ to a curve $d(t)$ in Q (which therefore has zero angular momentum) and then one acts on it by a time dependent group element solving the equation

$$\dot{g}(t) = g(t)\xi(t)$$

where $\xi(t) = \alpha(d(t))$, as in the theory of geometric phases — see Chapter 6 and Marsden, Montgomery and Ratiu [1990].

In general, one arrives at the reduced Hamiltonian description on $P_\mu \subset T^*(Q/G_\mu)$ with the amended potential by performing a Legendre transform in the non-degenerate variables; *i.e.*, the fiber variables corresponding to the fibers of $P_\mu \subset T^*(Q/G_\mu)$. For example, for abelian groups, one would perform a Legendre transformation in all the variables.

Remarks

1. This formulation of Lagrangian reduction appears to be useful for certain nonholonomic constraints (such as a nonuniform rigid body with a spherical shape rolling on a table); see Krishnaprasad, Yang, and Dayawansa [1991].

2. If one prefers, one can get a reduced Lagrangian description in the angular velocity rather than the angular momentum variables. In this approach, one keeps the relation $\xi = \alpha(q, v)$ unspecified till near the end. In this scenario, one starts by *enlarging* the space Q to $Q \times G$ (motivated by having a rotating frame in addition to the rotating structure, as in §3.7 below) and one adds to the given Lagrangian, the rotational energy for the G variables using the locked inertia tensor to form the kinetic energy — the motion on G is thus dependent on that on Q. In this description, one has ξ as an independent velocity variable and μ is its legendre transform. The Routhian is then seen already to be a Legendre transformation in the ξ and μ variables. One can delay making this Legendre transformation to the end, when the "locking device" that locks the motion on G to be that induced by the motion on Q by imposition of $\xi = \alpha(q, v)$ and $\xi = \mathbb{I}(q)^{-1}\mu$ or $\mathbf{J}(q, v) = \mu$.

3.7   Coupling to a Lie group

The following results are useful in the theory of coupling flexible structures to rigid bodies; see Krishnaprasad and Marsden [1987] and Simo, Posbergh and Marsden [1990].

Let G be a Lie group acting by canonical (Poisson) transformations on a Poisson manifold P. Define $\phi : T^*G \times P \to \mathfrak{g}^* \times P$ by

$$\phi(\alpha_g, x) = (TL_g^* \cdot \alpha_g, g^{-1} \cdot x) \tag{3.7.1}$$

where $g^{-1} \cdot x$ denotes the action of g^{-1} on $x \in P$. For example, if $G = SO(3)$ and α_g is a momentum variable which is given in coordinates on $T^*SO(3)$ by the momentum variables $p_\phi, p_\theta, p_\varphi$ conjugate to the Euler angles ϕ, θ, φ, then the mapping ϕ transforms α_g to body representation and transforms $x \in P$ to $g^{-1} \cdot x$, which represents x relative to the body.

For $F, K : \mathfrak{g}^* \times P \to \mathbb{R}$, let $\{F, K\}_-$ stand for the minus Lie-Poisson bracket holding the P variable fixed and let $\{F, K\}_P$ stand for the Poisson bracket on P with the variable $\mu \in \mathfrak{g}^*$ held fixed.

Endow $\mathfrak{g}^* \times P$ with the following bracket:

$$\{F, K\} = \{F, K\}_- + \{F, K\}_P - \mathbf{d}_x F \cdot \left(\frac{\delta K}{\delta \mu}\right)_P + \mathbf{d}_x K \cdot \left(\frac{\delta F}{\delta \mu}\right)_P \tag{3.7.2}$$

where $\mathbf{d}_x F$ means the differential of F with respect to $x \in P$ and the evaluation point (μ, x) has been suppressed.

Proposition 3.9 *The bracket (3.7.2) makes $\mathfrak{g}^* \times P$ into a Poisson manifold and $\phi : T^*G \times P \to \mathfrak{g}^* \times P$ is a Poisson map, where the Poisson structure on $T^*G \times P$ is given by the sum of the canonical bracket on T^*G and the bracket on P. Moreover, ϕ is G invariant and induces a Poisson diffeomorphism of $(T^*G \times P)/G$ with $\mathfrak{g}^* \times P$.*

Proof For $F, K : \mathfrak{g}^* \times P \to \mathbb{R}$, let $\bar{F} = F \circ \phi$ and $\bar{K} = K \circ \phi$. Then we want to show that $\{\bar{F}, \bar{K}\}_{T^*G} + \{\bar{F}, \bar{K}\}_P = \{F, K\} \circ \phi$. This will show ϕ is canonical. Since it is easy to check that ϕ is G invariant and gives a diffeomorphism of $(T^*G \times P)/G$ with $\mathfrak{g}^* \times P$, it follows that (3.7.2) represents the reduced bracket and so defines a Poisson structure.

To prove our claim, write $\phi = \phi_G \times \phi_P$. Since ϕ_G does not depend on x and the group action is assumed canonical, $\{\bar{F}, \bar{K}\}_P = \{F, K\}_P \circ \phi$. For the T^*G bracket, note that since ϕ_G is a Poisson map of T^*G to $\mathfrak{g}^*_-$, the terms involving ϕ_G will be $\{F, K\}_- \circ \phi$. The terms involving $\phi_P(\alpha_g, x) = g^{-1} \cdot x$ are given most easily by noting that the bracket of a function S of g with a function L of α_g is

$$\mathbf{d}_g S \cdot \frac{\delta L}{\delta \alpha_g}$$

where $\delta L / \delta \alpha_g$ means the fiber derivative of L regarded as a vector at g. This is paired with the covector $\mathbf{d}_g S$.

Letting $\Psi_x(g) = g^{-1} \cdot x$, we find by use of the chain rule that missing terms in the bracket are

$$\mathbf{d}_x F \cdot T\Psi_x \cdot \frac{\delta K}{\delta \mu} - \mathbf{d}_x K \cdot T\Psi_x \cdot \frac{\delta F}{\delta \mu}.$$

However, $T\Psi_x \cdot \dfrac{\delta K}{\delta \mu} = -\left(\dfrac{\delta K}{\delta \mu}\right)_P \circ \Psi_x$, so the preceding expression reduces to the last two terms in equation (3.7.2). $\blacksquare$

Suppose that the action of G on P has an Ad^* equivariant momentum map $\mathbf{J} : P \to \mathfrak{g}^*$. Consider the map $\alpha : \mathfrak{g}^* \times P \to \mathfrak{g}^* \times P$ given by

$$\alpha(\mu, x) = (\mu + \mathbf{J}(x), x). \tag{3.7.3}$$

Let the bracket $\{\,,\}_0$ on $\mathfrak{g}^* \times P$ be defined by

$$\{F, K\}_0 = \{F, K\}_- + \{F, K\}_P. \tag{3.7.4}$$

Thus $\{\,,\}_0$ is (3.7.2) with the coupling or interaction terms dropped. We claim that the map α eliminates the coupling:

Proposition 3.10 *The mapping* $\alpha : (\mathfrak{g}^* \times P, \{,\}) \to (\mathfrak{g}^* \times P, \{,\}_0)$ *is a Poisson diffeomorphism.*

Proof For $F, K : \mathfrak{g}^* \times P \to \mathbb{R}$, let $\hat{F} = F \circ \alpha$ and $\hat{K} = K \circ \alpha$. Letting $\nu = \mu + \mathbf{J}(x)$, and dropping evaluation points, we conclude that

$$\frac{\delta \hat{F}}{\delta \mu} = \frac{\delta F}{\delta \nu} \quad \text{and} \quad \mathbf{d}_x \hat{F} = \left\langle \frac{\delta F}{\delta \nu}, \mathbf{d}_x \mathbf{J} \right\rangle + \mathbf{d}_x F.$$

Substituting into the bracket (3.7.2), we get

$$\begin{aligned}
\{\hat{F}, \hat{K}\} = {}& -\left\langle \mu, \left[\frac{\delta F}{\delta \nu}, \frac{\delta K}{\delta \nu} \right] \right\rangle + \{F, K\}_P \\
& + \left\{ \left\langle \frac{\delta F}{\delta \nu}, \mathbf{d}_x \mathbf{J} \right\rangle, \left\langle \frac{\delta K}{\delta \nu}, \mathbf{d}_x \mathbf{J} \right\rangle \right\}_P \\
& + \left\{ \left\langle \frac{\delta F}{\delta \nu}, \mathbf{d}_x \mathbf{J} \right\rangle, \mathbf{d}_x K \right\}_P + \left\{ \mathbf{d}_x F, \left\langle \frac{\delta K}{\delta \nu}, \mathbf{d}_x \mathbf{J} \right\rangle \right\}_P \\
& - \left\langle \frac{\delta F}{\delta \nu}, \mathbf{d}_x \mathbf{J} \cdot \left(\frac{\delta K}{\delta \nu} \right)_P \right\rangle - \mathbf{d}_x F \cdot \left(\frac{\delta K}{\delta \nu} \right)_P \\
& + \left\langle \frac{\delta K}{\delta \nu}, \mathbf{d}_x \mathbf{J} \cdot \left(\frac{\delta F}{\delta \nu} \right)_P \right\rangle + \mathbf{d}_x K \cdot \left(\frac{\delta F}{\delta \nu} \right)_P. \quad (3.7.5)
\end{aligned}$$

Here, $\{\mathbf{d}_x F, \langle \delta K / \delta \nu, \mathbf{d}_x \mathbf{J} \rangle\}_P$ means the pairing of $\mathbf{d}_x F$ with the Hamiltonian vector field associated with the one form $\langle \delta K / \delta \nu, \mathbf{d}_x \mathbf{J} \rangle$, which is $(\delta K / \delta \nu)_P$, by definition of the momentum map. Thus the corresponding four terms in (3.7.5) cancel. Let us consider the remaining terms. First of all, we consider

$$\left\{ \left\langle \frac{\delta F}{\delta \nu}, \mathbf{d}_x \mathbf{J} \right\rangle, \left\langle \frac{\delta K}{\delta \nu}, \mathbf{d}_x \mathbf{J} \right\rangle \right\}_P. \quad (3.7.6)$$

Since $\mathbf{J}$ is equivariant, it is a Poisson map to $\mathfrak{g}^*_+$. Thus, (3.7.6) becomes $\langle \mathbf{J}, [\delta F / \delta \nu, \delta K / \delta \nu] \rangle$. Similarly each of the terms

$$-\left\langle \frac{\delta F}{\delta \nu}, \mathbf{d}_x \mathbf{J} \cdot \left(\frac{\delta K}{\delta \nu} \right)_P \right\rangle \quad \text{and} \quad \left\langle \frac{\delta K}{\delta \nu}, \mathbf{d}_x \mathbf{J} \cdot \left(\frac{\delta F}{\delta \nu} \right)_P \right\rangle$$

equal $-\langle \mathbf{J}, [\delta F / \delta \nu, \delta K / \delta \nu] \rangle$, and therefore these three terms collapse to $-\langle \mathbf{J}, [\delta F / \delta \nu, \delta K / \delta \nu] \rangle$ which combines with $-\langle \mu, [\delta F / \delta \nu, \delta K / \delta \nu] \rangle$ to produce the expression $-\langle \nu, [\delta F / \delta \nu, \delta K / \delta \nu] \rangle = \{F, K\}_-$. Thus, (3.7.5) collapses to (3.7.4). ∎

Remark This result is analogous to the isomorphism between the "Sternberg" and "Weinstein" representations of a reduced principal bundle. See

Sternberg [1977], Weinstein [1978], Montgomery, Marsden and Ratiu [1984] and Montgomery [1984].

Corollary 3.1 *Suppose $C(\nu)$ is a Casimir function on $\mathfrak{g}^*$. Then*

$$C(\mu, x) = C(\mu + \mathbf{J}(x))$$

is a Casimir function on $\mathfrak{g}^* \times P$ for the bracket (3.7.2).

We conclude this section with some consequences. The first is a connection with semi-direct products. Namely, we notice that if $\mathfrak{h}$ is another Lie algebra and G acts on $\mathfrak{h}$, we can reduce $T^*G \times \mathfrak{h}^*$ by G.

Corollary 3.2 *Giving $T^*G \times \mathfrak{h}^*$ the sum of the canonical and the "$-$" Lie-Poisson structure on $\mathfrak{h}^*$, the reduced space $(T^*G \times \mathfrak{h}^*)/G$ is $\mathfrak{g}^* \times \mathfrak{h}^*$ with the bracket*

$$\{F, K\} = \{F, K\}_{\mathfrak{g}^*} + \{F, K\}_{\mathfrak{h}^*} - \mathbf{d}_\nu F \cdot \left(\frac{\delta K}{\delta \mu}\right)_{\mathfrak{h}^*} + \mathbf{d}_\nu K \cdot \left(\frac{\delta F}{\delta \mu}\right)_{\mathfrak{h}^*} \quad (3.7.7)$$

where $(\mu, \nu) \in \mathfrak{g}^* \times \mathfrak{h}^*$, which is the Lie-Poisson bracket for the semidirect product $\mathfrak{g} \textcircled{S} \mathfrak{h}$.

Here is another consequence which reproduces the symplectic form on T^*G written in body coordinates (Abraham and Marsden [1978, p. 315]). We phrase the result in terms of brackets.

Corollary 3.3 *The map of T^*G to $\mathfrak{g}^* \times G$ given by $\alpha_g \mapsto (TL_g^* \alpha_g, g)$ maps the canonical bracket to the following bracket on $\mathfrak{g}^* \times G$:*

$$\{F, K\} = \{F, K\}_- + \mathbf{d}_g F \cdot TL_g \frac{\delta K}{\delta \mu} - \mathbf{d}_g K \cdot TL_g \frac{\delta F}{\delta \mu} \quad (3.7.8)$$

where $\mu \in \mathfrak{g}^*$ and $g \in G$.

Proof For $F : \mathfrak{g}^* \times G \to \mathbb{R}$, let $\bar{F}(\alpha_g) = F(\mu, g)$ where $\mu = TL_g^* \alpha_g$. The canonical bracket of $\bar{F}$ and $\bar{K}$ will give the $(-)$ Lie-Poisson structure via the μ dependence. The remaining terms are

$$\left\langle \mathbf{d}_g \bar{F}, \frac{\delta \bar{K}}{\delta p} \right\rangle - \left\langle \mathbf{d}_g \bar{K}, \frac{\delta \bar{F}}{\delta p} \right\rangle$$

where $\delta \bar{F}/\delta p$ means the fiber derivative of $\bar{F}$ regarded as a vector field and $\mathbf{d}_g \bar{K}$ means the derivative holding μ fixed. Using the chain rule, one gets (3.7.8). ∎

In the same spirit, one gets the next corollary by using the previous corollary twice.

Corollary 3.4 *The reduced Poisson space $(T^*G \times T^*G)/G$ is identifiable with the Poisson manifold $\mathfrak{g}^* \times \mathfrak{g}^* \times G$, with the Poisson bracket*

$$
\begin{aligned}
\{F, K\}(\mu_1, \mu_2, g) \;=\;& \{F, K\}^-_{\mu_1} + \{F, K\}^-_{\mu_2} \\
& -\mathbf{d}_g F \cdot TR_g \cdot \frac{\delta K}{\delta \mu_1} + \mathbf{d}_g K \cdot TR_g \cdot \frac{\delta F}{\delta \mu_1} \\
& +\mathbf{d}_g F \cdot TL_g \cdot \frac{\delta K}{\delta \mu_2} - \mathbf{d}_g K \cdot TL_g \cdot \frac{\delta F}{\delta \mu_2} \quad (3.7.9)
\end{aligned}
$$

where $\{F, K\}^-_{\mu_1}$ is the "$-$" Lie-Poisson bracket with respect to the first variable μ_1, and similarly for $\{F, K\}^-_{\mu_2}$.

Chapter 4

Relative Equilibria

In this chapter we give a variety of equivalent variational characterizations of relative equilibria. Most of these are well known, going back in the mid to late 1800's to Liouville, Laplace, Jacobi, Tait and Thomson, and Poincaré, continuing to more recent times in Smale [1970] and Abraham and Marsden [1978]. Our purpose is to assemble all these conveniently and to set the stage for the energy-momentum method in the next chapter. We begin with relative equilibria in the context of symplectic G-spaces and then later pass to the setting of simple mechanical systems.

4.1 Relative Equilibria on Symplectic Manifolds

Let $(P, \Omega, G, \mathbf{J}, H)$ be a symplectic G-space.

Definition 4.1 *A point $z_e \in P$ is called a* **relative equilibrium** *if*

$$X_H(z_e) \in T_{z_e}(G \cdot z_e)$$

i.e., if the Hamiltonian vector field at z_e points in the direction of the group orbit through z_e.

Theorem 4.1 Relative Equilibrium Theorem *Let $z_e \in P$ and let $z_e(t)$ be the dynamic orbit of X_H with $z_e(0) = z_e$ and let $\mu = \mathbf{J}(z_e)$. The following assertions are equivalent:*

77

  i z_e *is a relative equilibrium*

  ii $z_e(t) \in G_\mu \cdot z_e \subset G \cdot z_e$

 iii *there is a* $\xi \in \mathfrak{g}$ *such that* $z_e(t) = \exp(t\xi) \cdot z_e$

 iv *there is a* $\xi \in \mathfrak{g}$ *such that* z_e *is a critical point of the* **augmented Hamiltonian**

$$H_\xi(z) := H(z) - \langle \mathbf{J}(z) - \mu, \xi \rangle \tag{4.1.1}$$

  v z_e *is a critical point of the* **energy-momentum map**
      $H \times \mathbf{J} : P \to \mathbb{R} \times \mathfrak{g}^*$,

 vi z_e *is a critical point of* $H \mid \mathbf{J}^{-1}(\mu)$

 vii z_e *is a critical point of* $H \mid \mathbf{J}^{-1}(\mathcal{O})$, *where* $\mathcal{O} = G \cdot \mu \in \mathfrak{g}^*$

viii $[z_e] \in P_\mu$ *is a critical point of the reduced Hamiltonian* H_μ.

Remarks

1. In bifurcation theory one sometimes refers to relative equilibria as *rotating waves* and $G \cdot z_e$ as a *critical group orbit*.

2. Our definition is designed for use with a continuous group. If G were discrete, we would replace the above definition by the existence of a nontrivial subgroup $G_e \subset G$ such that $G_e \cdot z_e \subset \{z_e(t) \mid t \in \mathbb{R}\}$. In this case we would use the terminology *discrete relative equilibria* to distinguish it from the continuous case we treat.

3. The equivalence of **i** – **v** is general. However, the equivalence with **vi** – **viii** requires μ to be a regular value of $\mathbf{J}$ and for **vii** that the reduced manifold be nonsingular. It would be interesting to generalize the methods here to the singular case by using the results on bifurcation of momentum maps of Arms, Marsden and Moncrief [1981]. Roughly, one should use $\mathbf{J}$ alone for the singular part of $H^{-1}(\mu_e)$ (done by the Liapunov-Schmidt technique) and H_ξ for the regular part.

4. One can view **iv** as a constrained optimality criterion with ξ as a Lagrange multiplier.

5. We note that the criteria here for relative equilibria are related to the principle of symmetric criticality. See Palais [1979, 1985] for details.
   ◆

Proof The logic will go as follows:

$$\mathbf{i} \Rightarrow \mathbf{iv} \Rightarrow \mathbf{iii} \Rightarrow \mathbf{ii} \Rightarrow \mathbf{i}$$

$$\text{and} \quad \mathbf{iv} \Rightarrow \mathbf{v} \Rightarrow \mathbf{vi} \Rightarrow \mathbf{vii} \Rightarrow \mathbf{viii} \Rightarrow \mathbf{ii}.$$

First assume **i**, so $X_H(z_e) = \xi_P(z_e)$ for some $\xi \in \mathfrak{g}$. By definition of momentum map, this gives $X_H(z_e) = X_{\langle \mathbf{J}, \xi \rangle}(z_e)$ or $H_{H - \langle \mathbf{J}, \xi \rangle}(z_e) = 0$. Since

P is symplectic, this implies $H - \langle \mathbf{J}, \xi \rangle$ has a critical point at z_e, *i.e.*, that H_ξ has a critical point at z_e, which is **iv**.

Next, assume **iv**. Let φ_t denote the flow of X_H and ψ_t^ξ that of $X_{\langle \mathbf{J}, \xi \rangle}$, so $\psi_t^\xi(z) = \exp(t\xi) \cdot z$. Since H is G-invariant, φ_t and ψ_t^ξ commute, so the flow of $X_{H - \langle \mathbf{J}, \xi \rangle}$ is $\varphi_t \circ \psi_{-t}^\xi$. Since $H - \langle \mathbf{J}, \xi \rangle$ has a critical point at z_e, it is fixed by $\varphi_t \circ \psi_{-t}^\xi$, so $\varphi_t(\exp(-t\xi) \cdot z_e) = z_e$ for all $t \in \mathbb{R}$. Thus, $\varphi_t(z) = \exp(t\xi) \cdot z_e$, which is **iii**.

Condition **iii** shows that $z_e(t) \in G \cdot z_e$; but $z_e(t) \in \mathbf{J}^{-1}(\mu)$ and $G \cdot z_e \cap \mathbf{J}^{-1}(\mu) = G_\mu \cdot z_e$ by equivariance, so **iii** implies **ii** and by taking tangents, we see that **ii** implies **i**.

Assume **iv** again and notice that $H_\xi \mid \mathbf{J}^{-1}(\mu) = H \mid \mathbf{J}^{-1}(\mu)$, so **v** clearly holds. That **v** is equivalent to **vi** is one version of the Lagrange multiplier theorem.

Note by G-equivariance of $\mathbf{J}$ that $\mathbf{J}^{-1}(\mathcal{O})$ is the G-orbit of $\mathbf{J}^{-1}(\mu)$, so **vi** is equivalent to **vii** by G-invariance of H.

That **vi** implies **viii** follows by G-invariance of H and passing to the quotient.

Finally, **viii** implies that the equivalence class $[z_e]$ is a fixed point for the reduced dynamics, and so the orbit $z_e(t)$ in $\mathbf{J}^{-1}(\mu)$ projects to $[z_e]$; but this is exactly **ii**. ∎

To indicate the dependence of ξ on z_e when necessary, let us write $\xi(z_e)$. For the case $G = SO(3), \xi(z_e)$ is the *angular velocity of the uniformly rotating state* z_e.

Proposition 4.1 *Let z_e be a relative equilibrium. Then*

  i   *so is $g \cdot z_e$ for any $g \in G$ and $\xi(g z_e) = Ad_g[\xi(z_e)]$ and*

  ii $\xi(z_e) \in \mathfrak{g}_\mu$ *i.e.,* $Ad_{\exp t\xi}^* \mu = \mu$.

Proof Since

$$z_e(t) = \exp(t\xi) \cdot z_e \in \mathbf{J}^{-1}(\mu) \cap G \cdot z_e = G_\mu \cdot z_e,$$

$\exp(t\xi) \in G_\mu$, which is **ii**. Property **i** follows from the identity $H_{Ad_g \xi}(gz) = H_\xi(z)$. ∎

For $G = SO(3)$, **ii** means that $\xi(z_e)$ and μ are parallel vectors.

4.2   Cotangent Relative Equilibria

We now refine the relative equilibrium theorem to take advantage of the special context of simple mechanical systems. First notice that if H is of the form kinetic plus potential, then H_ξ can be rewritten as follows

$$H_\xi(z) = K_\xi(z) + V_\xi(q) + \langle \mu, \xi \rangle \qquad (4.2.1)$$

where $z = (q, p)$,

$$K_\xi(q, p) = \frac{1}{2} \| p - \mathbb{F}L(\xi_Q(q)) \|^2, \qquad (4.2.2)$$

and where

$$V_\xi(q) = V(q) - \frac{1}{2} \langle \xi, \mathbb{I}(q)\xi \rangle. \qquad (4.2.3)$$

Indeed,

$$\frac{1}{2} \| p - \mathbb{F}L(\xi_Q(q)) \|^2 + V(q) - \frac{1}{2} \langle \xi, \mathbb{I}(q)\xi \rangle + \langle \mu, \xi \rangle$$

$$= \frac{1}{2} \| p \|^2 - \langle\langle p, \mathbb{F}L(\xi_Q(q)) \rangle\rangle_q + \frac{1}{2} \| \mathbb{F}L(\xi_Q(q)) \|^2$$

$$\quad + V(q) - \frac{1}{2} \langle\langle \xi_Q(q), \xi_Q(q) \rangle\rangle_q + \langle \mu, \xi \rangle$$

$$= \frac{1}{2} \| p \|^2 - \langle p, \xi_Q(q) \rangle + \frac{1}{2} \langle\langle \xi_Q(q), \xi_Q(q) \rangle\rangle_q$$

$$\quad + V(q) - \frac{1}{2} \langle\langle \xi_Q(q), \xi_Q(q) \rangle\rangle_q + \langle \mu, \xi \rangle$$

$$= \frac{1}{2} \| p \|^2 - \langle \mathbf{J}(q, p), \xi \rangle + V(q) + \langle \mu, \xi \rangle$$

$$= H(q, p) - \langle \mathbf{J}(q, p) - \mu, \xi \rangle = H_\xi(q, p).$$

This calculation together with parts **i** and **iv** of the relative equilibrium theorem proves the following.

Proposition 4.2 *A point $z_e = (q_e, p_e)$ is a relative equilibrium if and only if there is a $\xi \in \mathfrak{g}$ such that*

i $p_e = \mathbb{F}L(\xi_Q(q_e))$ *and*
ii q_e *is a critical point of V_ξ.*

The functions K_ξ and V_ξ are called the **augmented** kinetic and potential energies respectively. The main point of this proposition is that *it reduces the job of finding relative equilibria to finding critical points of V_ξ.*

There is another interesting way to rearrange the terms in H_ξ, using the mechanical connection α and the amended potential V_μ. In carrying this out, it will be useful to note these two identities:

$$\mathbf{J}(\mathbb{F}L(\eta_Q(q))) = \mathbb{I}(q)\eta \qquad (4.2.4)$$

for all $\eta \in \mathfrak{g}$ and

$$\mathbf{J}(\alpha_\mu(q)) = \mu. \tag{4.2.5}$$

Indeed,

$$
\begin{aligned}
\langle \mathbf{J}(\mathbb{F}L(\eta_Q(q))), \chi \rangle &= \langle \mathbb{F}L(\eta_Q(q)), \chi_Q(q) \rangle \\
&= \langle\langle \eta_Q(q), \chi_Q(q) \rangle\rangle_q = \langle \mathbb{I}(q)\eta, \chi \rangle
\end{aligned}
$$

which gives (4.2.4) and (4.2.5) was proved in the last chapter.

At a relative equilibrium, the relation $p_e = \mathbb{F}L(\xi_Q(q_e))$ and (4.2.4) give

$$\mathbb{I}(q_e)\xi = \mathbf{J}(\mathbb{F}L(\xi_Q(q_e))) = \mathbf{J}(q_e, p_e) = \mu$$

i.e.,

$$\mu = \mathbb{I}(q_e)\xi \tag{4.2.6}$$

We now show that if $\xi = \mathbb{I}(q)^{-1}\mu$, then

$$H_\xi(z) = K_\mu(z) + V_\mu(q) \tag{4.2.7}$$

where

$$K_\mu(z) = \frac{1}{2}\|p - \alpha_\mu(q)\|^2 \tag{4.2.8}$$

and $V_\mu(q)$ is the amended potential as before. Indeed,

$$
\begin{aligned}
&K_\mu(z) + V_\mu(q) \\
&= \frac{1}{2}\|p - \alpha_\mu(q)\|^2 + V(q) + \frac{1}{2}\langle \mu, \mathbb{I}(q)^{-1}\mu \rangle \\
&= \frac{1}{2}\|p\|^2 - \langle\langle p, \alpha_\mu(q) \rangle\rangle_q + \frac{1}{2}\|\alpha_\mu(q)\|_q^2 + V(q) + \frac{1}{2}\langle \mu, \xi \rangle \\
&= \frac{1}{2}\|p\|^2 - \langle \mu, \alpha(\mathbb{F}L^{-1}(q, p)) \rangle + \frac{1}{2}\langle \mu, \alpha(\mathbb{F}L^{-1}(\alpha_\mu(q))) \rangle + V(q) + \frac{1}{2}\langle \mu, \xi \rangle \\
&= \frac{1}{2}\|p\|^2 - \langle \mu, \mathbb{I}^{-1}\mathbf{J}(q, p) \rangle + \frac{1}{2}\langle \mu, \mathbb{I}(q)^{-1}\mathbf{J}(\alpha_\mu(q)) \rangle + V(q) + \frac{1}{2}\langle \mu, \xi \rangle \\
&= \frac{1}{2}\|p\|^2 - \langle \mathbb{I}^{-1}\mu, \mathbf{J}(q, p) \rangle + \frac{1}{2}\langle \mu, \xi \rangle + V(q) + \frac{1}{2}\langle \mu, \xi \rangle \\
&= \frac{1}{2}\|p\|^2 + V(q) - \langle \mathbf{J}(q, p), \xi \rangle + \langle \mu, \xi \rangle = H_\xi(q, p)
\end{aligned}
$$

using $\mathbf{J}(\alpha_\mu(q)) = \mu$ and $\mathbf{J}(z) = \mu$. Next we observe that

$$p_e = \alpha_\mu(q_e) \tag{4.2.9}$$

since

$$\begin{aligned}
\langle \alpha_\mu(q_e), v \rangle &= \langle \mu, \mathbb{I}^{-1}(q_e)\mathbf{J}(\mathbb{F}L(v)) = \langle \mathbf{J}(q_e, p_e), \mathbb{I}^{-1}(q_e)\mathbf{J}(\mathbb{F}L(v)) \rangle \\
&= \langle p_e, [\mathbb{I}^{-1}(q_e)\mathbf{J}(\mathbb{F}L(v))]_Q \rangle \\
&= \langle \mathbb{F}L(\xi_Q(q)), [\mathbb{I}^{-1}(q_e)\mathbf{J}(\mathbb{F}L(v))]_Q \rangle \\
&= \langle \xi, \mathbf{J}(\mathbb{F}L(v)) \rangle = \langle \xi_Q(q_e), \mathbb{F}L(v) \rangle \\
&= \langle\langle \xi_Q(q_e), v \rangle\rangle_q = \langle \mathbb{F}L(\xi_Q(q_e)), v \rangle = \langle p_e, v \rangle.
\end{aligned}$$

These calculations prove the following result:

Proposition 4.3 *A point $z_e = (q_e, p_e)$ is a relative equilibrium if and only if*

  i $p_e = \alpha_\mu(q_e)$ *and*
  ii q_e *is a critical point of V_μ.*

One can also derive (4.2.9) from (4.2.7) and also one can get **ii** directly from **ii** of Proposition 4.2. As we shall see later, it will be V_μ that gives the sharpest stability results, as opposed to V_ξ.

In Simo, Lewis and Marsden [1991] it is shown how, in an appropriate sense, V_ξ and V_μ are related by a Legendre transformation. We also note that this result follows directly from the method of Lagrangian reduction given in §3.6.

Proposition 4.4 *If $z_e = (q_e, p_e)$ is a relative equilibrium and $\mu = \mathbf{J}(z_e)$, then μ is an equilibrium for the Lie-Poisson system on $\mathfrak{g}^*$ with Hamiltonian $h(\nu) = \frac{1}{2}\langle \nu, \mathbb{I}(q_e^{-1})\nu \rangle$, i.e., the locked inertia Hamiltonian.*

Proof The Hamiltonian vector field of h is

$$X_h(\nu) = ad^*_{\delta h/\delta \nu}\nu = ad^*_{\mathbb{I}(q_e)^{-1}\nu}\nu.$$

However, $\mathbb{I}(q_e)^{-1}\mu = \xi$ by (4.2.4) and $ad^*_\xi \mu = 0$ by Proposition 4.1, part **ii**. Hence $X_h(\mu) = 0$. ∎

4.3   Examples

Example 1 *The spherical pendulum* Recall from Chapter 3 that V_ξ and V_μ are given by

$$V_\xi(\theta) = -mgR\cos\theta - \frac{\xi^2}{2}mR^2\sin^2\theta$$

and

$$V_\mu(\theta) = -mgR\cos\theta + \frac{1}{2}\frac{\mu^2}{mR^2\sin^2\theta}.$$

The relation (4.2.6) is $\mu = mR^2\sin^2\theta_e\xi$, the usual relation between angular momentum μ and angular velocity ξ. The conditions of Proposition 4.2 are:

$$(p_\theta)_e = 0, \quad (p_\varphi)_e = mR^2\sin\theta_e\xi$$

and $V_\xi'(\theta_e) = 0$, *i.e.*,

$$mgR\sin\theta_e - \xi^2 mR^2\sin\theta_e\cos\theta_e = 0$$

i.e., $\sin\theta_e = 0$ or $g = R^2\xi^2\cos\theta_e$. Thus, the relative equilibria correspond to the pendulum in the upright or down position (singular case) or, with $\theta_e \neq 0, \pi$ to any $\theta_e \neq \pi/2$ with

$$\xi = \pm\sqrt{\frac{g}{R\cos\theta_e}}.$$

Using the relation between μ and ξ, the reader can check that Proposition 4.3 gives the same result.

Example 2 *The double spherical pendulum* In Chapter 3 we saw that

$$V_\mu(\mathbf{q}_1, \mathbf{q}_2) = m_1 g \mathbf{q}_1 \cdot \mathbf{k} + m_2 g(\mathbf{q}_1 + \mathbf{q}_2) \cdot \mathbf{k} + \frac{1}{2}\frac{\mu^2}{I}, \qquad (4.3.1)$$

where

$$I(\mathbf{q}_1, \mathbf{q}_2) = m_1\|\mathbf{q}_1^\perp\|^2 + m_2\|\mathbf{q}_1^\perp + \mathbf{q}_2^\perp\|^2.$$

The relative equilibria are computed by finding the critical points of V_μ. There are *. . .* obvious relative equilibria — the ones with $\mathbf{q}_1^\perp = 0$ and $\mathbf{q}_2^\perp = 0$, in which the individual pendula are pointing vertically upwards or vertically downwards. We now search for solutions with each pendulum pointing downwards, and with $\mathbf{q}_1^\perp \neq 0$ and $\mathbf{q}_2^\perp \neq 0$. There are some other relative equilibria with one of the pendula pointing upwards, as we shall discuss below.

We next express V_μ as a function of $\mathbf{q}_1^\perp$ and $\mathbf{q}_2^\perp$ by using the constraints, which gives the third components:

$$q_1^3 = -\sqrt{l_1^2 - \|\mathbf{q}_1^\perp\|^2} \quad \text{and} \quad q_2^3 = -\sqrt{l_2^2 - \|\mathbf{q}_2^\perp\|^2}.$$

Thus,

$$V_\mu(\mathbf{q}_1^\perp, \mathbf{q}_2^\perp) = -(m_1 + m_2)\sqrt{l_1^2 - \|\mathbf{q}_1^\perp\|^2} - m_2\sqrt{l_2^2 - \|\mathbf{q}_2^\perp\|^2} + \frac{1}{2}\frac{\mu^2}{I}. \quad (4.3.2)$$

Setting the derivatives of V_μ equal to zero gives

$$(m_1 + m_2)g\frac{\mathbf{q}_1^\perp}{\sqrt{l_1^2 - \|\mathbf{q}_1^\perp\|^2}} = \frac{\mu^2}{I^2}[(m_1 + m_2)\mathbf{q}_1^\perp + m_2\mathbf{q}_2^\perp]$$

$$m_2 g\frac{\mathbf{q}_2^\perp}{\sqrt{l_2^2 - \|\mathbf{q}_2^\perp\|^2}} = \frac{\mu^2}{I^2}[m_2(\mathbf{q}_1^\perp + \mathbf{q}_2^\perp)] \qquad (4.3.3)$$

We note that the equations for critical points of V_ξ give the same equations with $\mu = I\xi$.

From (4.3.3) we see that the vectors $\mathbf{q}_1^\perp$ and $\mathbf{q}_2^\perp$ are parallel. Therefore, define a parameter α by

$$\mathbf{q}_2^\perp = \alpha\mathbf{q}_1^\perp \qquad (4.3.4)$$

and let λ be defined by

$$\|\mathbf{q}_1^\perp\| = \lambda l_1. \qquad (4.3.5)$$

Notice that α and λ determine the *shape* of the relative equilibrium. Define the *system parameters* r and m by

$$r = \frac{l_2}{l_1}, \quad m = \frac{m_1 + m_2}{m_2}, \qquad (4.3.6)$$

so that conditions (4.3.3) are equivalent to

$$\left.\begin{array}{c} \dfrac{mg}{l_1}\dfrac{1}{\sqrt{1 - \lambda^2}} = \dfrac{\mu^2}{I^2}(m + \alpha) \\[2mm] \dfrac{g}{l_1}\dfrac{\alpha}{\sqrt{r^2 - \alpha^2\lambda^2}} = \dfrac{\mu^2}{I^2}(1 + \alpha) \end{array}\right\} \qquad (4.3.7)$$

The restrictions on the parameters are as follows: First, from $\|\mathbf{q}_1^\perp\| \leq l_1$ and $\|\mathbf{q}_2^\perp\| \leq l_2$ we get

$$0 \leq \lambda \leq \max\{r/\alpha, 1\} \qquad (4.3.8)$$

and next, from the equations (4.3.7) we get the restriction that either

$$\alpha > 0 \quad \text{or} \quad -m < \alpha < -1. \qquad (4.3.9)$$

The intervals $(-\infty, m)$ and $(-1, 0)$ are also possible and correspond to pendulum configurations with the first and second pendulum inverted, respectively. Dividing the equations (4.3.7) to eliminate μ and using a little algebra proves the following result.

Theorem 4.2 *All of the relative equilibria of the double spherical pendulum apart the four equilibria with the two pendula vertical are given by the points on the graph of*

$$\lambda^2 = \frac{L^2 - r^2}{L^2 - \alpha^2} \qquad (4.3.10)$$

where

$$L(\alpha) = \left(1 + \frac{\alpha}{m}\right)\left(\frac{\alpha}{1+\alpha}\right),$$

subject to the restriction $0 \le \lambda^2 \le r^2/\alpha^2$.

From (4.3.7) we can express either μ or ξ in terms of α. In Figures 4.3.1 and 4.3.2 we show the relative equilibria for two sample values of the system parameters. Note that there is a bifurcation of relative equilibria for fixed m and increasing r, and that it occurs within the range of restricted values of α and λ. Also note that there can be two or three relative equilibria for a given set of system parameters.

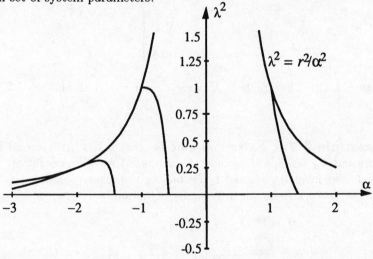

Figure 4.3.1: The graphs of λ^2 verses α for $r = 1, m = 2$ and of $\lambda^2 = r^2/\alpha^2$.

The bifurcation of relative equilibria that happens between Figures 4.3.1 and 4.3.2 does so along the curve in the (r, m) plane given by

$$r = \frac{2m}{1+m}$$

as is readily seen. For instance, for $m = 2$ one gets $r = 4/3$, in agreement with the figures. The above results are in agreement with those of Baillieul [1987].

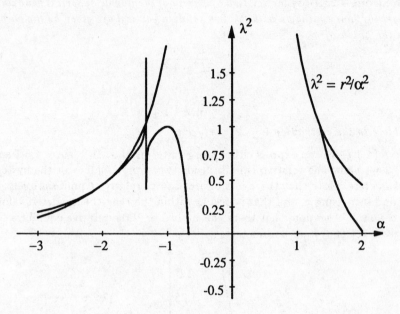

Figure 4.3.2: The graph of λ^2 verses α for $r = 1.35$ and $m = 2$ and of $\lambda^2 = r^2/\alpha^2$.

Example 3 *The water molecule* Because it is a little complicated to work out V_μ (except at symmetric molecules) we will concentrate on the conditions involving H_ξ and V_ξ for finding the relative equilibria.

Let us work these out in turn. By definition,

$$
\begin{aligned}
H_\xi &= H - \langle \mathbf{J}, \xi \rangle \\
&= \frac{1}{4m\overline{M}}\|\sigma\|^2 + \frac{\|\pi\|^2}{m} + \frac{1}{2\mathcal{M}}\|\mathbf{j}\|^2 + V \qquad (4.3.11) \\
&\quad - \pi \cdot (\xi \times \mathbf{r}) - \sigma \cdot (\xi \times \mathbf{s}). \qquad (4.3.12)
\end{aligned}
$$

Thus, the conditions for a critical point associated with $\delta H_\xi = 0$ are:

$$
\delta\pi \quad : \quad \frac{2}{m}\pi = \xi \times \mathbf{r} \qquad (4.3.13)
$$

$$
\delta\sigma \quad : \quad \frac{1}{2m\overline{M}}\sigma = \xi \times \mathbf{s} \qquad (4.3.14)
$$

$$
\delta\mathbf{r} \quad : \quad \frac{\partial V}{\partial \mathbf{r}} = \pi \times \xi \qquad (4.3.15)
$$

$$\delta \mathbf{s} \quad : \quad \frac{\partial V}{\partial \mathbf{s}} = \sigma \times \xi. \tag{4.3.16}$$

The first two equations express π and σ in terms of $\mathbf{r}$ and $\mathbf{s}$. When they are substituted in the second equations, we get conditions on $(\mathbf{r}, \mathbf{s})$ alone:

$$\frac{\partial V}{\partial \mathbf{r}} = \frac{m}{2}(\xi \times \mathbf{r}) \times \xi = \frac{m}{2}\left[\xi(\xi \cdot \mathbf{r}) - \mathbf{r}\|\xi\|^2\right] \tag{4.3.17}$$

and

$$\frac{\partial V}{\partial \mathbf{s}} = 2m\overline{M}(\xi \times \mathbf{s}) \times \xi = 2m\overline{M}\left[\xi(\xi \cdot \mathbf{s}) - \mathbf{s}\|\xi\|^2\right]. \tag{4.3.18}$$

These are the conditions for a relative equilibria we sought. Any collection $(\xi, \mathbf{r}, \mathbf{s})$ satisfying (4.3.17) and (4.3.18) gives a relative equilibrium.

Now consider V_ξ for the water molecule:

$$
\begin{aligned}
V_\xi(\mathbf{r}, \mathbf{s}) \;=\; & V(\mathbf{r}, \mathbf{s}) - \frac{1}{2}\left[\|\xi\|^2\left(\frac{m}{2}\|\mathbf{r}\|^2 + 2M\overline{m}\|\mathbf{s}\|^2\right)\right] \\
& - \frac{1}{2}\left[\frac{m}{2}(\xi \cdot \mathbf{r})^2 + 2M\overline{m}(\xi \cdot \mathbf{s})^2\right].
\end{aligned}
\tag{4.3.19}
$$

The conditions for a critical point are:

$$\delta \mathbf{r} \quad : \quad \frac{\partial V}{\partial \mathbf{r}} - \|\xi\|^2 \frac{m\mathbf{r}}{2} - \frac{m}{2}(\xi \cdot \mathbf{r})\xi = 0 \tag{4.3.20}$$

$$\delta \mathbf{s} \quad : \quad \frac{\partial V}{\partial \mathbf{s}} - \|\xi\|^2 \cdot 2M\overline{m}\mathbf{s} - 2M\overline{m}(\xi \cdot \mathbf{s})\xi = 0, \tag{4.3.21}$$

which coincide with (4.3.17) and (4.3.18). The extra condition from Proposition 4.2, namely $p_e = \mathbb{F}L(\xi_Q(q_e))$, gives (4.3.13) and (4.3.14).

4.4 The Rigid Body

Since examples like the water molecule have both rigid and internal variables, it is important to understand our constructions for the rigid body itself.

The rotation group $SO(3)$ consists of all orthogonal linear transformations of Euclidean three space to itself, which have determinant one. Its Lie algebra, denoted $so(3)$, consists of all 3×3 skew matrices, which we identify with $\mathbb{R}^3$ by the isomorphism $\hat{\ } : \mathbb{R}^3 \to so(3)$ defined by

$$\Omega \mapsto \hat{\Omega} = \begin{bmatrix} 0 & -\Omega^3 & \Omega^2 \\ \Omega^3 & 0 & -\Omega^1 \\ -\Omega^2 & \Omega^1 & 0 \end{bmatrix}, \tag{4.4.1}$$

where $\Omega = (\Omega^1, \Omega^2, \Omega^3)$. One checks that for any vectors $\mathbf{r}$, and Θ,

$$\hat{\Omega}\mathbf{r} = \Omega \times \mathbf{r}, \quad \text{and} \quad \hat{\Omega}\hat{\Theta} - \hat{\Theta}\hat{\Omega} = (\Omega \times \Theta)\hat{.} \tag{4.4.2}$$

Equations (4.4.1) and (4.4.2) identify the Lie algebra $so(3)$ with $\mathbb{R}^3$ and the Lie algebra bracket with the cross product of vectors. If $\Lambda \in SO(3)$ and $\Omega \in so(3)$, the **adjoint action** is given by

$$[\Lambda\Theta]\hat{} = \text{Ad}_\Lambda\hat{\Theta}. \tag{4.4.3}$$

The fact that the adjoint action is a Lie algebra homomorphism, corresponds to the identity

$$\Lambda(\mathbf{r} \times \mathbf{s}) = \Lambda\mathbf{r} \times \Lambda\mathbf{s}, \tag{4.4.4}$$

for all $\mathbf{r}, \mathbf{s} \in \mathbb{R}^3$.

Given $\Lambda \in SO(3)$, let $\hat{\mathbf{v}}_\Lambda$ denote an element of the tangent space to $SO(3)$ at Λ. Since $SO(3)$ is a submanifold of $GL(3)$, the general linear group, we can identify $\hat{\mathbf{v}}_\Lambda$ with a 3×3 matrix, which we denote with the same letter. Linearizing the defining (submersive) condition $\Lambda\Lambda^T = 1$ gives

$$\Lambda\hat{\mathbf{v}}_\Lambda^T + \hat{\mathbf{v}}_\Lambda\Lambda^T = 0, \tag{4.4.5}$$

which defines $T_\Lambda SO(3)$. We can identify $T_\Lambda SO(3)$ with $so(3)$ by two isomorphisms: First, given $\hat{\Theta} \in so(3)$ and $\Lambda \in SO(3)$, define $(\Lambda, \hat{\Theta}) \mapsto \hat{\Theta}_\Lambda \in T_\Lambda SO(3)$ by letting $\hat{\Theta}_\Lambda$ be the left invariant extension of $\hat{\Theta}$:

$$\hat{\Theta}_\Lambda := T_e L_\Lambda \cdot \hat{\Theta} \cong (\Lambda, \Lambda\hat{\Theta}). \tag{4.4.6}$$

Second, given $\hat{\theta} \in so(3)$ and $\Lambda \in SO(3)$, define $(\Lambda, \hat{\theta}) \mapsto \hat{\theta}_\Lambda \in T_\Lambda SO(3)$ through right translations by setting

$$\hat{\theta}_\Lambda := T_e R_\Lambda \cdot \hat{\theta} \cong (\Lambda, \hat{\theta}\Lambda). \tag{4.4.7}$$

The notation is partially dictated by continuum mechanics considerations; upper case letters are used for the body (or convective) variables and lower case for the spatial (or Eulerian) variables. Often, the base point is omitted and with an abuse of notation we write $\Lambda\hat{\Theta}$ and $\hat{\theta}\Lambda$ for $\hat{\Theta}_\Lambda$ and $\hat{\theta}_\Lambda$, respectively.

The dual space to $so(3)$ is identified with $\mathbb{R}^3$ using the **standard dot product**: $\Pi \cdot \Theta = \frac{1}{2}\text{tr}[\hat{\Pi}^T\hat{\Theta}]$. This extends to the **left-invariant pairing** on $T_\Lambda SO(3)$ given by

$$\langle \hat{\Pi}_\Lambda, \hat{\Theta}_\Lambda \rangle = \frac{1}{2}\text{tr}[\hat{\Pi}_\Lambda^T\hat{\Theta}_\Lambda] = \frac{1}{2}\text{tr}[\hat{\Pi}^T\hat{\Theta}] = \Pi \cdot \mathbf{w}. \tag{4.4.8}$$

Write elements of $so(3)^*$ as $\hat{\Pi}$, where $\Pi \in \mathbb{R}^3$, (or $\hat{\pi}$ with $\pi \in \mathbb{R}^3$) and elements of $T_\Lambda^* SO(3)$ as

$$\hat{\Pi}_\Lambda = (\Lambda, \Lambda\hat{\Pi}), \tag{4.4.9}$$

for the body representation, and for the spatial representation

$$\hat{\pi}_\Lambda = (\Lambda, \hat{\pi}\Lambda). \tag{4.4.10}$$

Again, explicit indication of the base point will often be omitted and we shall simply write $\Lambda\hat{\Pi}$ and $\hat{\pi}\Lambda$ for $\hat{\Pi}_\Lambda$ and $\hat{\pi}_\Lambda$, respectively. If (4.4.9) and (4.4.10) represent the same covector, then

$$\hat{\pi} = \Lambda\hat{\Pi}\Lambda^T, \tag{4.4.11}$$

which coincides with the co-adjoint action. Equivalently, using the isomorphism (4.4.2) we have

$$\pi = \Lambda\Pi. \tag{4.4.12}$$

The mechanical set-up for rigid body dynamics is as follows: the configuration manifold Q and the phase space P are

$$Q = SO(3) \quad \text{and} \quad P = T^* SO(3) \tag{4.4.13}$$

with the canonical symplectic structure.

The Lagrangian for the free rigid body is, as in Chapter 1, its kinetic energy:

$$L(\Lambda, \dot{\Lambda}) = \frac{1}{2} \int_{\mathcal{B}} \rho_{\text{ref}}(X) \|\dot{\Lambda}X\|^2 d^3 X \tag{4.4.14}$$

where ρ_{ref} is a mass density on the reference configuration $\mathcal{B}$. The Lagrangian is evidently left invariant. The corresponding metric is the left invariant metric given at the identity by

$$\langle\langle\hat{\xi}, \hat{\eta}\rangle\rangle = \int_{\mathcal{B}} \rho_{\text{ref}}(\mathbf{X})\hat{\xi}(\mathbf{X}) \cdot \hat{\eta}(\mathbf{X}) d^3\mathbf{X}$$

or on $\mathbb{R}^3$ by

$$\langle\langle\mathbf{a}, \mathbf{b}\rangle\rangle = \int_{\mathcal{B}} \rho_{\text{ref}}(\mathbf{X})(\mathbf{a} \times \mathbf{X}) \cdot (\mathbf{b} \times \mathbf{X}) d^3 X = \mathbf{a} \cdot \mathbb{J}\mathbf{b} \tag{4.4.15}$$

where

$$\mathbb{J} = \int_{\mathcal{B}} \rho_{\text{ref}}(\mathbf{X})[\|\mathbf{X}\|^2 \mathbf{I} - \mathbf{X} \otimes \mathbf{X}] d^3\mathbf{X} \tag{4.4.16}$$

is the *inertia tensor*.

The corresponding Hamiltonian is

$$H = \frac{1}{2}\Pi \cdot \mathbb{J}^{-1}\Pi = \frac{1}{2}\pi \cdot \mathbb{I}^{-1}\pi \qquad (4.4.17)$$

where $\mathbb{I} = \Lambda \mathbb{J} \Lambda^{-1}$ is the *time dependent inertia tensor*.

The expression $H = \frac{1}{2}\Pi \cdot \mathbb{J}^{-1}\Pi$ reflects the *left invariance* of H under the action of $SO(3)$. Thus *left reduction* by $SO(3)$ to *body coordinates* induces a function on the quotient space $T^*SO(3)/SO(3) \cong so(3)^*$. The symplectic leaves are spheres, $\|\Pi\| = $ constant. The induced function on these spheres is given by (4.4.17) regarded as a function of Π. The dynamics on this sphere is obtained by intersection of the sphere $\|\Pi\|^2 = $ constant and the ellipsoid $H = $ constant that we discussed in Chapter 1.

Consistent with the preceding discussion, let $G = SO(3)$ act from the *left* on $Q = SO(3)$ *i.e.,*

$$Q \cdot \Lambda = L_Q \Lambda = Q\Lambda, \qquad (4.4.18)$$

for all $\Lambda \in SO(3)$ and $Q \in G$. Hence the action of $G = SO(3)$ on $P = T^*SO(3)$ is by cotangent lift of left translations. Since the infinitesimal generator associated with $\hat{\xi} \in so(3)$ is obtained as

$$\hat{\xi}_{SO(3)}(\Lambda) = \frac{d}{dt}\exp[t\hat{\xi}]\Lambda \mid_{t=0} = \hat{\xi}\Lambda, \qquad (4.4.19)$$

the corresponding momentum map is

$$\mathbf{J}(\hat{\xi})(\hat{\pi}_\Lambda) = -\frac{1}{2}\mathrm{tr}[\hat{\pi}_\Lambda^T \xi_{so(3)}\Lambda] = -\frac{1}{2}\mathrm{tr}[\Lambda^T \hat{\pi}^T \hat{\xi}\Lambda] = -\frac{1}{2}\mathrm{tr}[\hat{\pi}^T\hat{\xi}] = \pi \cdot \xi, \qquad (4.4.20)$$

i.e.,

$$\mathbf{J}(\hat{\pi}_\Lambda) = \hat{\pi}, \quad \text{or} \quad \mathbf{J}(\hat{\xi}) = \pi \cdot \xi. \qquad (4.4.21)$$

To locate the relative equilibria, consider

$$H_\xi = H - [\mathbf{J}(\xi) - \pi_e \cdot \xi] = \frac{1}{2}\pi \cdot \mathbb{I}^{-1}\pi - \xi \cdot (\pi - \pi_e). \qquad (4.4.22)$$

To find its critical points, we note that although $\hat{\pi}_\Lambda \in T^*SO(3)$ are the basic variables, it is more convenient to regard H_ξ as a function of $(\Lambda, \pi) \in SO(3) \times \mathbb{R}^{3*}$ through the isomorphism (4.4.2).

Thus, let $\hat{\pi}_{\Lambda_e} \cong (\Lambda_e, \hat{\pi}_e \Lambda_e) \in T^*SO(3)$ be a relative equilibrium point. For any $\delta\theta \in \mathbb{R}^3$ we construct the curve

$$\epsilon \mapsto \Lambda_\epsilon = \exp[\epsilon\widehat{\delta\theta}]\Lambda_e \in SO(3), \qquad (4.4.23)$$

which starts at Λ_e and

$$\left.\frac{d}{d\epsilon}\Lambda_\epsilon\right|_{\epsilon=0} = \widehat{\delta\theta}\Lambda. \qquad (4.4.24)$$

Let $\delta\pi \in (\mathbb{R}^3)^*$ and consider the curve in $(\mathbb{R}^3)^*$ defined as

$$\epsilon \mapsto \pi_\epsilon = \pi_e + \epsilon\delta\pi \in (\mathbb{R}^3)^*, \qquad (4.4.25)$$

which starts at π_e. These constructions induce a curve $\epsilon \mapsto \hat{\pi}_{\Lambda_e} \in T^*SO(3)$ via the isomorphism (4.4.10); that is, $\hat{\pi}_{\Lambda_e} := (\Lambda_\epsilon, \hat{\pi}_\epsilon\Lambda_\epsilon)$. With this notation at hand we compute the first variation using the chain rule. Let

$$\delta H_\xi \mid_e := \left.\frac{d}{d\epsilon}H_{\xi,\epsilon}\right|_{\epsilon=0} = 0, \qquad (4.4.26)$$

where

$$H_{\xi,\epsilon} := \frac{1}{2}\pi_\epsilon \cdot \mathbb{I}_\epsilon^{-1}\pi_\epsilon - \xi \cdot \pi_\epsilon \quad \text{and} \quad \mathbb{I}_\epsilon^{-1} := \Lambda_\epsilon\mathbb{J}^{-1}\Lambda_\epsilon^T.$$

In addition, at equilibrium, $\mathbf{J}(z_e) = \mu$ reads

$$\pi = \pi_e. \qquad (4.4.27)$$

To compute δH_ξ, observe that

$$
\begin{aligned}
\left.\frac{1}{2}\pi_e \cdot \frac{d}{d\epsilon}\mathbb{I}_\epsilon^{-1}\pi_e\right|_{\epsilon=0} &= \frac{1}{2}[\pi_e \cdot \widehat{\delta\theta}\mathbb{I}_e^{-1} - \mathbb{I}_e^{-1}\widehat{\delta\theta}]\pi_e \\
&= \frac{1}{2}[\pi_e \cdot \delta\theta \times \mathbb{I}_e^{-1}\pi_e - \mathbb{I}_e^{-1}\pi_e \cdot \delta\theta \times \pi_e] \\
&= \delta\theta \cdot (\mathbb{I}_e^{-1}\pi_e \times \pi_e), \qquad (4.4.28)
\end{aligned}
$$

where we have made use of elementary vector product identities. Thus,

$$\delta H_\xi \mid_e = \delta\pi \cdot [\mathbb{I}_e^{-1}\pi_e - \xi] + \delta\theta \cdot [\mathbb{I}_e^{-1}\pi_e \times \pi_e] = 0. \qquad (4.4.29)$$

From this relation we obtain the two equilibrium conditions:

$$\mathbb{I}_e^{-1}\pi_e \times \pi_e = 0, \quad \text{and} \quad \mathbb{I}_e^{-1}\pi_e = \xi. \qquad (4.4.30)$$

Equivalently,

$$\xi \times \pi_e = 0, \quad \text{and} \quad \mathbb{I}_e^{-1}\xi = \lambda\xi, \qquad (4.4.31)$$

where $\lambda > 0$ by positive definiteness of $\mathbb{I}_e = \Lambda_e\mathbb{J}\Lambda_e^T$. These conditions state that π_e is aligned with a principal axis, and that the rotation is about this axis. Note that $\pi = \mathbb{I}_e\omega_e$, where $\mathbb{I}_e$ is the spatial inertial dyadic and $\pi_e = \mathbb{I}_e\xi$, so that ξ does indeed correspond to the angular velocity.

Equivalently, we can locate the relative equilibria using Proposition 4.2. The locked inertia tensor (as the notation suggests) is just $\mathbb{I}$ and here $p_e = \mathbb{F}L(\xi_Q(q_e))$ reads $\pi_e = \mathbb{I}_e\xi$, while $q_e = \Lambda_e$ is required to be a critical point of V_ξ or V_μ, where $\mu = \pi_e = \mathbb{I}\xi$. Thus

$$V_\xi(\Lambda) = \frac{1}{2}\xi \cdot (\mathbb{I}\xi) \quad \text{and} \quad V_\mu(\Lambda) = -\frac{1}{2}\mu \cdot (\mathbb{I}^{-1}\mu). \tag{4.4.32}$$

The calculation giving (4.4.28) shows that Λ_e is a critical point of V_ξ, or of V_μ iff $\xi \times \pi_e = 0$.

These calculations show that rigid body relative equilibria correspond to steady rotational motions about their principal axes. The "global" point of view taken above is perhaps a bit long winded, but it is a point of view that is useful in the long run.

Chapter 5

The Energy-Momentum Method

This chapter develops the energy-momentum method of Simo, Posbergh and Marsden [1990, 1991] and Simo, Lewis and Marsden [1991]. This is a technique for determining the stability of relative equilibria and for putting the equations of motion linearized at a relative equilibrium, into normal form. This normal form is based on a special decomposition of variations into rigid and internal components that gives a block structure to the Hamiltonian and symplectic structure. There has been considerable development of stability and bifurcation techniques over the last decade, and some properties like block diagonalization have been seen in a variety of problems; for example, this appears to be what is happening in Morrison and Pfirsch [1990].

5.1 The General Technique

In these lectures we will confine ourselves to the *regular case*; that is, we assume z_e is a relative equilibrium that is also a regular point (*i.e.*, $\mathfrak{g}_{z_e} = \{0\}$, or z_e has a *discrete* isotropy group) and $\mu = \mathbf{J}(z_e)$ is a *generic point* in $\mathfrak{g}^*$ (*i.e.*, its orbit is of maximal dimension). We are seeking conditions for stability of z_e modulo G_μ. To do so, find a subspace

$$\mathcal{S} \subset T_{z_e} P$$

satisfying two conditions (see Figure 5.1.1):

i $\mathcal{S} \subset \ker \mathbf{DJ}(z_e)$ and

ii $\mathcal{S}$ is transverse to the G_μ-orbit within $\ker \mathbf{DJ}(z_e)$.

The *energy-momentum method* is as follows:

a   find $\xi \in \mathfrak{g}$ such that $\delta H_\xi(z_e) = 0$

b   test $\delta^2 H_\xi(z_e)$ for definiteness on $\mathcal{S}$.

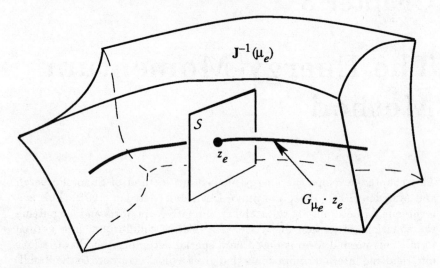

Figure 5.1.1: The energy-momentum method tests the second variation of the augmented Hamiltonian for definiteness on the space $\mathcal{S}$.

Theorem 5.1 The Energy-Momentum Theorem. *If $\delta^2 H_\xi(z_e)$ is definite, then z_e is G_μ-orbitally stable in $\mathbf{J}^{-1}(\mu)$ and G-orbitally stable in P.*

Proof First, one has G_μ-orbital stability within $\mathbf{J}^{-1}(\mu)$ because the reduced dynamics induces a well defined dynamics on the orbit space P_μ; this dynamics induces a dynamical system on $\mathcal{S}$ for which H_ξ is an invariant function. Since it has a non-degenerate extremum at z_e, invariance of its level sets gives the required stability.

Second, one gets G_ν orbital stability within $\mathbf{J}^{-1}(\nu)$ for ν close to μ since the form of Figure 5.1.1 changes in a regular way for nearby level sets, as μ is both a regular value and a generic point in $\mathfrak{g}^*$. ∎

For results for non-generic μ or non-regular μ, see Patrick [1990] and Lewis [1991].

A well known example that may be treated as an instance of the energy-momentum method, where the necessity of considering orbital stability is especially clear, is the stability of solitons in the KdV equation. (A result of Benjamin [1972] and Bona [1974].) Here *one gets stability of solitons modulo translations*; one cannot expect literal stability because a slight change in the amplitude cases a translational drift, but the soliton *shape* remains dynamically stable.

This example is actually infinite dimensional, and here one must employ additional hypotheses for Theorem 5.1 to be valid. Two possible infinite dimensional versions are as follows: one uses convexity hypotheses going back to Arnold [1969] (see Holm et al. [1985] for a concise summary) or, what is appropriate for the KdV equation, employ Sobolev spaces on which the calculus argument of Theorem 5.1 is correct — one needs the energy norm defined by $\delta^2 H_\xi(z_e)$ to be equivalent to a Sobolev norm in which one has global existence theorems and on which H_ξ is a smooth function. Sometimes, such as in three dimensional elasticity, *this can be a serious difficulty* (see, for instance, Ball and Marsden [1984]). In other situations, a special analysis is needed, as in Wan and Pulvirente [1984] and Batt [1990].

The energy-momentum method can be compared to Arnold's energy-Casimir method that takes place on the Poisson manifold P/G rather than on P itself. Consider the diagram in Figure 5.1.2.

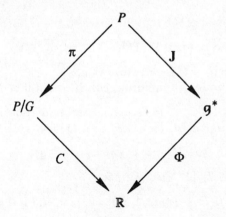

Figure 5.1.2: Comparing the energy-Casimir and energy-momentum methods.

The **energy-Casimir method** searches for a Casimir function C on P/G such that $H + C$ has a critical point at $u_e = [z_e]$ for the reduced system and then requires definiteness of $\delta^2(H+C)(u_e)$. If one has a relative equilibrium z_e (with its associated μ and ξ), and one can find an invariant function Φ on $\mathfrak{g}^*$ such that

$$\frac{\delta\Phi}{\delta\mu}(\mu) = -\xi,$$

then one can define a Casimir C by $C \circ \pi = \Phi \circ \mathbf{J}$ and $H + C$ will have a critical point at u_e.

Remark This diagram gives another way of viewing symplectic reduction, namely, as the level sets of Casimir functions in P/G. In general, the symplectic reduced spaces may be viewed as the symplectic leaves in P/G. These may or may not be realizable as level sets of Casimir functions. ◆

The energy-momentum method is more powerful in that it does not depend on the existence of invariant functions, which is a serious difficulty in some examples, such as geometrically exact elastic rods, certain plasma problems, and 3-dimensional ideal flow. On the other hand, the energy-Casimir method is able to treat some singular cases rather easily since P/G can still be a smooth manifold in the singular case. However, there is still the problem of interpretation of the results in P; see Patrick [1990].

For simple mechanical systems, one way to choose S is as follows. Let

$$\mathcal{V} = \{\delta q \in T_{q_e}Q \mid \langle\langle \delta q, \chi_Q(q_e)\rangle\rangle = 0 \quad \text{for all} \quad \chi \in \mathfrak{g}_\mu\},$$

the *metric* orthogonal of the tangent space to the G_μ-orbit in Q. Let

$$S = \{\delta z \in \ker \mathbf{DJ}(z_e) \mid T\pi_Q \cdot \delta z \in \mathcal{V}\}$$

where $\pi_Q : T^*Q = P \to Q$ is the projection.

The following *gauge invariance condition* will be helpful.

Proposition 5.1 *Let $z_e \in P$ be a relative equilibrium, and let $G \cdot z_e = \{g \cdot z_e \mid g \in G\}$ be the orbit through z_e with tangent space*

$$T_{z_e}(G \cdot z_e) = \{\eta_P(z_e) \mid \eta \in \mathfrak{g}\}. \tag{5.1.1}$$

Then, for any $\delta z \in T_{z_e}[\mathbf{J}^{-1}(\mu)]$, we have

$$\delta^2 H_\xi(z_e) \cdot (\eta_P(z_e), \delta z) = 0 \quad \text{for all} \quad \eta \in \mathfrak{g}. \tag{5.1.2}$$

Proof Since $H : P \to \mathbb{R}$ is G-invariant, Ad^*-equivariance of the momentum map gives

$$\begin{aligned} H_\xi(g \cdot z) &= H(g \cdot z) - \langle \mathbf{J}(g \cdot z), \xi\rangle + \langle\mu, \xi\rangle \\ &= H(z) - \langle Ad^*_{g^{-1}}(\mathbf{J}(z)), \xi\rangle + \langle\mu, \xi\rangle \\ &= H(z) - \langle \mathbf{J}(z), Ad_{g^{-1}}(\xi)\rangle + \langle\mu, \xi\rangle, \tag{5.1.3} \end{aligned}$$

for any $g \in G$ and $z \in P$. Choosing $g = \exp(t\eta)$ with $\eta \in \mathfrak{g}$, and differentiating (5.1.3) with respect to t gives

$$\mathbf{d}H_\xi(z) \cdot \eta_P(z) = -\left\langle \mathbf{J}(z), \frac{d}{dt}\bigg|_{t=0} Ad_{\exp(-t\eta)}(\xi) \right\rangle = \langle \mathbf{J}(z), [\eta, \xi] \rangle. \quad (5.1.4)$$

Taking the variation of (5.1.4) with respect to z, evaluating at z_e and using the fact that $\mathbf{d}H_\xi(z_e) = 0$, one gets the expression

$$\delta^2 H_\xi(z_e)(\eta_P(z_e), \delta z) = \langle T_{z_e}\mathbf{J} \cdot \delta z, [\eta, \xi] \rangle, \quad (5.1.5)$$

which vanishes if $T_{z_e}\mathbf{J}(z_e) \cdot \delta z = 0$, *i.e.*, if $\delta z \in \ker[T_{z_e}\mathbf{J}(z_e)] = T_{z_e}\mathbf{J}^{-1}(\mu_e)$.
∎

In particular, from the above result and Proposition 4.1 we have

Proposition 5.2 $\delta^2 H_\xi(z_e)$ *vanishes identically on* $\ker[T_{z_e}\mathbf{J}(z_e)]$ *along the directions tangent to the orbit* $G_\mu \cdot z_e$; *that is*

$$\delta^2 H_\xi(z_e) \cdot (\eta_P(z_e), \zeta_P(z_e)) = 0 \quad \text{for any} \quad \eta, \zeta \in \mathfrak{g}_\mu. \quad (5.1.6)$$

Proof A general fact about reduction is that $T_{z_e}(G_\mu \cdot z_e) = T_{z_e}(G \cdot z_e) \cap \ker[T_{z_e}\mathbf{J}(z_e)]$. Since $T_{z_e}(G_\mu \cdot z_e) \subset T_{z_e}(G \cdot z_e)$ the result follows from (5.1.2) by taking $\delta z = \xi_P(z_e)$ with $\xi \in \mathfrak{g}_\mu$. ∎

These propositions confirm the general geometric picture set out in Figure 5.1.1. They show explicitly that the orbit directions are neutral directions of $\delta^2 H_\xi(z_e)$, so one has no chance of proving definiteness except on a transverse to the G_μ-orbit.

Remark Failure to understand these simple distinctions between stability and orbital stability can sometimes lead to misguided assertions like "Arnold's method can only work at symmetric states". (See Andrews [1984], Chern and Marsden [1989] and Carnevale and Shepard [1989].) ♦

5.2 Example: The Rigid Body

The classical result about free rigid body stability mentioned in Chapter 1 is that uniform rotation about the longest and shortest principal axes are stable motions, while motion about the intermediate axis is unstable. The simplest way to see this is by using the energy-Casimir method.

We begin with the equations of motion in body representation:

$$\dot{\Pi} = \frac{d\Pi}{dt} = \Pi \times \Omega \tag{5.2.1}$$

where $\Pi, \Omega \in \mathbb{R}^3$, Ω is the angular velocity, and Π is the angular momentum. Assuming principal axis coordinates, $\Pi_j = I_j \Omega_j$ for $j = 1, 2, 3$, where $I = (I_1, I_2, I_3)$ is the diagonalized moment of inertia tensor, $I_1, I_2, I_3 > 0$. As we saw in Chapter 2, this system is Hamiltonian in the Lie-Poisson structure with the kinetic energy Hamiltonian

$$H(\Pi) = \frac{1}{2}\Pi \cdot \Omega = \frac{1}{2}\sum_{i=0}^{3} \frac{\Pi_i^2}{I_i}, \tag{5.2.2}$$

and that for a smooth function $\Phi : \mathbb{R} \to \mathbb{R}$, the function

$$C_\Phi(\Pi) = \Phi\left(\frac{1}{2}\|\Pi\|^2\right) \tag{5.2.3}$$

is a Casimir function. We first choose Φ such that $H_{C_\Phi} := H + C_\Phi$ has a critical point at a given equilibrium point of (5.2.1). Such points occur when Π is parallel to Ω. We can assume that the equilibrium solution is $\Pi_e = (1, 0, 0)$. The derivative of

$$H_{C_\Phi}(\Pi) = \frac{1}{2}\sum_{i=0}^{3} \frac{\Pi_i^2}{I_i} + \Phi\left(\frac{1}{2}\|\Pi\|^2\right)$$

is

$$\mathbf{D}H_{C_\Phi}(\Pi) \cdot \delta\Pi = \left(\Omega + \Pi\Phi'\left(\frac{1}{2}\|\Pi\|^2\right)\right) \cdot \delta\Pi. \tag{5.2.4}$$

This equals zero at $\Pi = (1, 0, 0)$, provided that

$$\Phi'\left(\frac{1}{2}\right) = -\frac{1}{I_1}. \tag{5.2.5}$$

The second derivative at the equilibrium $\Pi_e = (1, 0, 0)$ is

$$\mathbf{D}^2 H_{C_\Phi}(\Pi_e) \cdot (\delta\Pi, \delta\Pi)$$

$$= \delta\Omega \cdot \delta\Pi + \Phi'\left(\frac{1}{2}\|\Pi_e\|^2\right)\|\delta\Pi\|^2 + (\Pi_e \cdot \delta\Pi)^2 \Phi''\left(\frac{1}{2}\|\Pi_e\|^2\right)$$

$$= \sum_{i=0}^{3} \frac{(\delta\Pi_i)^2}{I_i} - \frac{\|\delta\Pi\|^2}{I_1} + \Phi''\left(\frac{1}{2}\right)(\delta\Pi_1)^2 \tag{5.2.6}$$

$$= \left(\frac{1}{I_2} - \frac{1}{I_1}\right)(\delta\Pi_2)^2 + \left(\frac{1}{I_3} - \frac{1}{I_1}\right)(\delta\Pi_3)^2 + \Phi''\left(\frac{1}{2}\right)(\delta\Pi_1)^2.$$

This quadratic form is positive definite if and only if

$$\Phi'' \left(\frac{1}{2} \right) > 0 \tag{5.2.7}$$

and

$$I_1 > I_2, \qquad I_1 > I_3. \tag{5.2.8}$$

Consequently,

$$\Phi(x) = - \left(\frac{1}{I_1} \right) x + \left(x - \frac{1}{2} \right)^2$$

satisfies (5.2.5) and makes the second derivative of H_{C_Φ} at $(1,0,0)$ positive definite, so stationary rotation around the longest axis is stable. The quadratic form is negative definite provided

$$\Phi'' \left(\frac{1}{2} \right) < 0 \tag{5.2.9}$$

and

$$I_1 < I_2, \qquad I_1 < I_3. \tag{5.2.10}$$

A specific function Φ satisfying the requirements (5.2.5) and (5.2.9) is

$$\Phi(x) = - \left(\frac{1}{I_1} \right) x - \left(x - \frac{1}{2} \right)^2 .$$

This proves that the rigid body in steady rotation around the short axis is (Liapunov) stable. Finally, the quadratic form (5.2.6) is indefinite if

$$I_1 > I_2, \qquad I_3 > I_1. \tag{5.2.11}$$

or the other way around. One needs an additional argument to show that rotation around the middle axis is unstable. Perhaps the simplest way is as follows: Linearizing equation (5.2.1) at $\Pi_e = (1,0,0)$ yields the linear constant coefficient system

$$\frac{d}{dt} \delta \Pi = \delta \Pi \times \Omega_e + \Pi_e \times \delta \Omega = \left(0, \frac{I_3 - I_1}{I_3 I_1} \delta \Pi_3, \frac{I_1 - I_2}{I_1 I_2} \delta \Pi_2 \right)$$

$$= \begin{bmatrix} 0 & 0 & 0 \\ 0 & 0 & \dfrac{I_3 - I_1}{I_3 I_1} \\ 0 & \dfrac{I_1 - I_2}{I_1 I_2} & 0 \end{bmatrix} \delta \Pi. \tag{5.2.12}$$

On the tangent space at Π_e to the sphere of radius $\|\Pi_e\| = 1$, the linear operator given by this linearized vector field has matrix given by the lower right 2×2 block, whose eigenvalues are

$$\pm \frac{1}{I_1 \sqrt{I_2 I_3}} \sqrt{(I_1 - I_2)(I_3 - I_1)}.$$

Both eigenvalues are real by (5.2.11) and one is strictly positive. Thus Π_e is spectrally unstable and thus is (nonlinearly) unstable. Thus, *in the motion of a free rigid body, rotation around the long and short axes is (Liapunov) stable and around the middle axis is unstable*.

The energy-Casimir method deals with stability for the dynamics on Π-space. However, what one wants (corresponding to what one "sees") is orbital stability for the motion in $T^*SO(3)$. This follows from stability in the variable Π, but one has to use conservation of spatial angular momentum to see it. The energy-momentum method gives this right away. On the other hand, the energy-momentum method appears to be more complicated for the rigid body. Its utility is really only revealed when one does more complex examples, as in Simo, Posbergh and Marsden [1990] and Lewis and Simo [1990].

In §4.4 we set up the rigid body as a mechanical system on $T^*SO(3)$ and we located the relative equilibria. By differentiating as in (4.4.29) we find the following formula:

$$\delta^2 H_\xi \mid_e ((\delta\pi, \delta\theta), (\delta\pi, \delta\theta)) = \qquad\qquad (5.2.13)$$

$$[\delta\pi^T \quad \delta\theta^T] \begin{bmatrix} \mathbb{I}_e^{-1} & (\mathbb{I}_e^{-1} - \lambda\mathbf{1})\hat{\pi}_e \\ -\hat{\pi}_e(\mathbb{I}_e^{-1} - \lambda\mathbf{1}) & -\hat{\pi}_e(\mathbb{I}_e^{-1} - \lambda\mathbf{1})\hat{\pi}_e \end{bmatrix} \begin{bmatrix} \delta\pi \\ \delta\theta \end{bmatrix}.$$

Note that this matrix is 6×6. Corresponding to the two conditions on $\mathcal{S}$, we restrict the admissible variations $(\delta, \pi, \delta\theta) \in \mathbb{R}^{3*} \times \mathbb{R}^3$. Here, $\mathbf{J}(\hat{\pi}_\Lambda) = \hat{\pi}_\Lambda$; hence $\mu = \hat{\pi}_e$ and $T_{z_e}(G_\mu \cdot z_e) = $ infinitesimal rotations about the axis π_e; *i.e.*, multiples of π_e, or equivalently ξ. Variations that are orthogonal to this space and also lie in the space $\delta\pi = 0$ (which is the condition $\delta J = \delta\pi = 0$) are of the form $\delta\theta$ with $\delta\theta \perp \pi_e$. Thus, we choose

$$\mathcal{S} = \{(\delta\pi, \delta\theta) \mid \delta\pi = 0, \delta\theta \perp \pi_e\}. \qquad\qquad (5.2.14)$$

Note that $\delta\theta$ is a variation that rotates π_e on the sphere

$$\mathcal{O}_{\pi_e} := \{\pi \in \mathbb{R}^3 \mid \|\pi\|^2 = \|\pi_e\|^2\}, \qquad\qquad (5.2.15)$$

which is the co-adjoint orbit through π_e. The second variation (5.2.13) restricted to the subspace $\mathcal{S}$ is given by

$$\delta^2 H_{\xi,\mu}|_e = \delta\theta \cdot (\hat{\pi}_e^T (\mathbb{I}_e^{-1} - \lambda \mathbf{1}) \hat{\pi}_e) \delta\theta = (\pi_e \times \delta\theta) \cdot (\mathbb{I}_e^{-1} - \lambda \mathbf{1}) (\pi_e \times \delta\theta). \quad (5.2.16)$$

If λ is the largest or smallest eigenvalue of $\mathbb{I}$, (5.2.16) will be definite; note that the null space of $\mathbb{I}^{-1} - \lambda\mathbf{1}$ in (5.2.16) consists of vectors parallel to π_e, which have been excluded. Also note that in this example, $\mathcal{S}$ is a 2-dimensional space and (5.2.16) in fact represents a 2×2 matrix. As we shall see below, this 2×2 block can also be viewed as $\delta^2 V_\mu(q_e)$ on the space of rigid variations.

5.3   Block Diagonalization

If the energy-momentum method is applied to mechanical systems with Hamiltonian H of the form kinetic energy (K) plus potential (V), it is possible to choose variables in a way that makes the determination of stability conditions sharper and more computable. In this set of variables (with the conservation of momentum constraint and a gauge symmetry constraint imposed on $\mathcal{S}$), the second variation of $\delta^2 H_\xi$ block diagonalizes; schematically,

$$\delta^2 H_\xi = \begin{bmatrix} \begin{bmatrix} 2 \times 2\,\text{rigid} \\ \text{body block} \end{bmatrix} & 0 \\ 0 & \begin{bmatrix} \text{Internal vibration} \\ \text{block} \end{bmatrix} \end{bmatrix}.$$

Furthermore, the internal vibrational block takes the form

$$\begin{bmatrix} \text{Internal vibration} \\ \text{block} \end{bmatrix} = \begin{bmatrix} \delta^2 V_\mu & 0 \\ 0 & \delta^2 K_\mu \end{bmatrix}$$

where V_μ is the amended potential defined earlier, and K_μ is a momentum shifted kinetic energy, so formal stability is equivalent to $\delta^2 V_\mu > 0$ and that the overall structure is stable when viewed as a rigid structure, which, as far as stability is concerned, separates the overall rigid body motions from the internal motions of the system under consideration.

The dynamics of the internal vibrations (such as the elastic wave speeds) depend on the rotational angular momentum. That is, the internal vibrational block is μ-dependent, but in a way we shall explicitly calculate. On the other hand, these two types of motions do not *dynamically decouple*, since the symplectic form does *not* block diagonalize. However, the symplectic form takes on a particularly simple *normal form* as we shall see.

This allows one to put the *linearized equations of motion* into normal form as well.

To define the rigid-internal splitting, we begin with a splitting in configuration space. Consider (at a relative equilibrium) the space $\mathcal{V}$ defined above as the metric orthogonal to $\mathfrak{g}_\mu(g)$ in $T_q Q$. Here we drop the subscript e for notational convenience. Then we split

$$\mathcal{V} = \mathcal{V}_{\text{RIG}} \oplus \mathcal{V}_{\text{INT}} \tag{5.3.1}$$

as follows. Define

$$\mathcal{V}_{\text{RIG}} = \{\eta_Q(q) \in T_q Q \mid \eta \in \mathfrak{g}_\mu^\perp\} \tag{5.3.2}$$

where $\mathfrak{g}_\mu^\perp$ is the orthogonal complement to $\mathfrak{g}_\mu$ in $\mathfrak{g}$ with respect to the locked inertia metric. (This choice of orthogonal complement depends on q, but we do not include this in the notation.) From (3.3.1) it is clear that $\mathcal{V}_{\text{RIG}} \subset \mathcal{V}$ and that $\mathcal{V}_{\text{RIG}}$ has the dimension of the coadjoint orbit through μ. Next, define

$$\mathcal{V}_{\text{INT}} = \{\delta q \in \mathcal{V} \mid \langle \eta, [\mathbf{D}\mathbb{I}(q) \cdot \delta q] \cdot \xi \rangle = 0 \quad \text{for all} \quad \eta \in \mathfrak{g}_\mu^\perp\} \tag{5.3.3}$$

where $\xi = \mathbb{I}(q)^{-1}\mu$. An equivalent definition is

$$\mathcal{V}_{\text{INT}} = \{\delta q \in \mathcal{V} \mid [\mathbf{D}\mathbb{I}(q)^{-1} \cdot \delta q] \cdot \mu \in \mathfrak{g}_\mu\}.$$

The definition of $\mathcal{V}_{\text{INT}}$ has an interesting mechanical interpretation in terms of the objectivity of the centrifugal force in case $G = SO(3)$; see Simo, Lewis and Marsden [1991].

Define the **Arnold form** $\mathcal{A}_\mu : \mathfrak{g}_\mu^\perp \times \mathfrak{g}_\mu^\perp \to \mathbb{R}$ by

$$\mathcal{A}_\mu(\eta, \zeta) = \langle ad_\eta^* \mu, \chi_{(q,\mu)}(\zeta) \rangle = \langle \mu, ad_\eta \chi_{(q,\mu)}(\zeta) \rangle, \tag{5.3.4}$$

where $\chi_{(q,\mu)} : \mathfrak{g}_\mu^\perp \to \mathfrak{g}$ is defined by

$$\chi_{(q,\mu)}(\zeta) = \mathbb{I}(q)^{-1} ad_\zeta^* \mu + ad_\zeta \mathbb{I}(q)^{-1} \mu.$$

The Arnold form appears in Arnold's [1966] stability analysis of relative equilibria in the special case $Q = G$. At a relative equilibrium, the form $\mathcal{A}_\mu$ is symmetric, as is verified either directly or by recognizing it as the second variation of V_μ on $\mathcal{V}_{\text{RIG}} \times \mathcal{V}_{\text{RIG}}$ (see (5.3.15) below for this calculation).

At a relative equilibrium, the form $\mathcal{A}_\mu$ is degenerate as a symmetric bilinear form on $\mathfrak{g}_\mu^\perp$ when there is a non-zero $\zeta \in \mathfrak{g}_\mu^\perp$ such that

$$\mathbb{I}(q)^{-1} ad_\zeta^* \mu + ad_\zeta \mathbb{I}(q)^{-1} \mu \in \mathfrak{g}_\mu;$$

in other words, when $\mathbb{I}(q)^{-1} : \mathfrak{g}^* \to \mathfrak{g}$ has a *nontrivial symmetry* relative to the coadjoint-adjoint action of $\mathfrak{g}$ (restricted to $\mathfrak{g}_\mu^\perp$) on the space of linear maps from $\mathfrak{g}^*$ to $\mathfrak{g}$. (When one is not at a relative equilibrium, we say the Arnold form is *non-degenerate* when $\mathcal{A}_\mu(\eta, \zeta) = 0$ for all $\eta \in \mathfrak{g}_\mu^\perp$ implies $\zeta = 0$.) This means, for $G = SO(3)$ that $\mathcal{A}_\mu$ is non-degenerate if μ is not in a multidimensional eigenspace of $\mathbb{I}^{-1}$. Thus, *if the locked body is not symmetric (i.e., a Lagrange top), then the Arnold form is non-degenerate.*

Proposition 5.3 *If the Arnold form is non-degenerate, then*

$$\mathcal{V} = \mathcal{V}_{\mathrm{RIG}} \oplus \mathcal{V}_{\mathrm{INT}}. \tag{5.3.5}$$

Indeed, non-degeneracy of the Arnold form implies $\mathcal{V}_{\mathrm{RIG}} \cap \mathcal{V}_{\mathrm{INT}} = \{0\}$ and, at least in the finite dimensional case, a dimension count gives (5.3.5). In the infinite dimensional case, the relevant ellipticity conditions are needed.

The split (5.3.5) can now be used to induce a split of the phase space

$$\mathcal{S} = \mathcal{S}_{\mathrm{RIG}} \oplus \mathcal{S}_{\mathrm{INT}}. \tag{5.3.6}$$

Using a more mechanical viewpoint, Simo, Lewis and Marsden [1991] show how $\mathcal{S}_{\mathrm{RIG}}$ can be defined by extending $\mathcal{V}_{\mathrm{RIG}}$ from positions to momenta using *superposed rigid motions*. For our purposes, the important characterization of $\mathcal{S}_{\mathrm{RIG}}$ is via the mechanical connection:

$$\mathcal{S}_{\mathrm{RIG}} = T_q \alpha_\mu \cdot \mathcal{V}_{\mathrm{RIG}} \tag{5.3.7}$$

so $\mathcal{S}_{\mathrm{RIG}}$ is isomorphic to $\mathcal{V}_{\mathrm{RIG}}$. Since α_μ maps Q to $\mathbf{J}^{-1}(\mu)$ and $\mathcal{V}_{\mathrm{RIG}} \subset \mathcal{V}$, we get $\mathcal{S}_{\mathrm{RIG}} \subset \mathcal{S}$. Define

$$\mathcal{S}_{\mathrm{INT}} = \{\delta z \in \mathcal{S} \mid \delta q \in \mathcal{V}_{\mathrm{INT}}\}; \tag{5.3.8}$$

then (5.3.6) holds if the Arnold Form is non-degenerate. Next, we write

$$\mathcal{S}_{\mathrm{INT}} = \mathcal{W}_{\mathrm{INT}} \oplus \mathcal{W}_{\mathrm{INT}}^*, \tag{5.3.9}$$

where $\mathcal{W}_{\mathrm{INT}}$ and $\mathcal{W}_{\mathrm{INT}}^*$ are defined as follows:

$$\mathcal{W}_{\mathrm{INT}} = T_q \alpha_\mu \cdot \mathcal{V}_{\mathrm{INT}} \quad \text{and} \quad \mathcal{W}_{\mathrm{INT}}^* = \{\mathrm{vert}(\gamma) \mid \gamma \in [\mathfrak{g} \cdot q]^0\} \tag{5.3.10}$$

where $\mathfrak{g} \cdot q = \{\zeta_Q(q) \mid \zeta \in \mathfrak{g}\}$, $[\mathfrak{g} \cdot q]^0 \subset T_q^* Q$ is its annihilator, and $\mathrm{vert}(\gamma) \in T_z(T^*Q)$ is the vertical lift of $\gamma \in T_q^* Q$; in coordinates, $\mathrm{vert}(q^i, \gamma_j) = (q^i, p_j, 0, \gamma_j)$. The vertical lift is given intrinsically by taking the tangent to the curve $\sigma(s) = z + s\gamma$ at $s = 0$.

Theorem 5.2 Block Diagonalization Theorem *In the splittings intro-duced above at a relative equilibrium, $\delta^2 H_\xi(z_e)$ and the symplectic form Ω_{z_e} have the following form:*

$$
\delta^2 H_\xi(z_e) = \left[\begin{array}{ccc} \left[\begin{array}{c} \text{Arnold} \\ \text{form} \end{array} \right] & 0 & 0 \\ 0 & \delta^2 V_\mu & 0 \\ 0 & 0 & \delta^2 K_\mu \end{array} \right]
$$

and

$$
\Omega_{z_e} = \left[\begin{array}{ccc} \left[\begin{array}{c} \text{coadjoint orbit} \\ \text{symplectic form} \end{array} \right] & \left[\begin{array}{c} \text{internal rigid} \\ \text{coupling} \end{array} \right] & 0 \\ -\left[\begin{array}{c} \text{internal rigid} \\ \text{coupling} \end{array} \right] & S & 1 \\ 0 & -1 & 0 \end{array} \right]
$$

where the columns represent elements of $S_{\mathrm{RIG}}, W_{\mathrm{INT}}$ and W_{INT}^*, respec-tively.

Remarks

1 For the viewpoint of these splittings being those of a connection, see Lewis, Marsden, Ratiu and Simo [1990].

2 Arms, Fischer and Marsden [1975] and Marsden [1981] describe an im-portant phase space splitting of $T_z P$ at a point $z \in \mathbf{J}^{-1}(0)$ into three pieces, namely

$$
T_z P = T_z(G \cdot z) \oplus S \oplus [T_z \mathbf{J}^{-1}(0)]^\perp,
$$

where S is, as above, a complement to the gauge piece $T_z(G \cdot z)$ within $T_z \mathbf{J}^{-1}(0)$ and can be taken to be the same as S defined in §5.1, and $[T_z \mathbf{J}^{-1}(0)]^\perp$ is a complement to $T_z \mathbf{J}^{-1}(0)$ within $T_z P$. This splitting has profound implications for general relativistic fields, where it is called *Mon-crief's splitting*. See, for example, Arms, Marsden and Moncrief [1982] and references therein. Thus, we get, all together, a six-way splitting of $T_z P$. It would be of interest to explore the geometry of this situation fur-ther and arrive at normal forms for the linearized dynamical equations valid on all of $T_z P$.

3 It is also of interest to link the normal forms here with those in singularity theory. In particular, can one use the forms here as first terms in higher order normal forms?   ◆

The terms appearing in the formula for Ω_{z_e} will be explained in §5.4 below. We now give some of the steps in the proof.

Lemma 5.1 $\delta^2 H_\xi(z_e) \cdot (\Delta z, \delta z) = 0$ for $\Delta z \in \mathcal{S}_{\mathrm{RIG}}$ and $\delta z \in \mathcal{S}_{\mathrm{INT}}$.

Proof In Chapter 3 we saw that for $z \in \mathbf{J}^{-1}(\mu)$,

$$H_\xi(z) = K_\mu(z) + V_\mu(q) \tag{5.3.11}$$

where each of the functions K_μ and V_μ has a critical point at the relative equilibrium. It suffices to show that each term separately satisfies the lemma. The second variation of K_μ is

$$\delta^2 K_\mu(z_e) \cdot (\Delta z, \delta z) = \langle\langle \Delta p - T\alpha_\mu(q_e) \cdot \Delta q, \delta p - T\alpha_\mu(q_e) \cdot \delta q \rangle\rangle. \tag{5.3.12}$$

By (5.3.7), $\Delta p = T\alpha_\mu(q_e) \cdot \Delta q$, so this expression vanishes. The second variation of V_μ is

$$\delta^2 V_\mu(q_e) \cdot (\Delta q, \delta q) = \delta(\delta V(q) \cdot \Delta q + \frac{1}{2}\mu[\mathbf{D}\mathbb{I}^{-1}(q) \cdot \Delta q]\mu) \cdot \delta q. \tag{5.3.13}$$

By G-invariance, $\delta V(q) \cdot \Delta q$ is zero for any $q \in Q$, so the first term vanishes. As for the second, let $\Delta q = \eta_Q(q)$ where $\eta \in \mathfrak{g}_\mu^\perp$. As was mentioned in Chapter 3, $\mathbb{I}$ is equivariant;

$$\langle \mathbb{I}(gq)Ad_g\chi, Ad_g\zeta \rangle = \langle \mathbb{I}(q)\chi, \zeta \rangle.$$

Differentiating this with respect to g at $g = e$ in the direction η gives

$$(\mathbf{D}\mathbb{I}(q) \cdot \Delta q)\chi + \mathbb{I}(q)ad_\eta\chi + ad_\eta^*[\mathbb{I}(q)\chi] = 0$$

and so

$$\begin{aligned}
\mathbf{D}\mathbb{I}^{-1}(q) \cdot \Delta q\, \nu &= -\mathbb{I}(q)^{-1}\mathbf{D}\mathbb{I}(q) \cdot \Delta q\mathbb{I}(q)^{-1}\nu \\
&= ad_\eta\mathbb{I}(q)^{-1}\nu + \mathbb{I}(q)^{-1}ad_\eta^*\nu. \tag{5.3.14}
\end{aligned}$$

Differentiating with respect to q in the direction δq and letting $\nu = \mu$ gives

$$\mathbf{D}^2\mathbb{I}^{-1}(q) \cdot (\Delta q, \delta q)] \cdot \mu = ad_\eta[(D\mathbb{I}(q)^{-1} \cdot \delta q)\mu] + (D\mathbb{I}(q)^{-1} \cdot \delta q) \cdot ad_\eta^*\mu.$$

Pairing with μ and using symmetry of $\mathbb{I}$ gives

$$\langle \mu, D^2\mathbb{I}^{-1}(q) \cdot (\Delta q, \delta q) \cdot \mu \rangle = 2\langle \mu, ad_\eta[D\mathbb{I}(q)^{-1} \cdot \delta q)\mu] \rangle. \tag{5.3.15}$$

If δq is a rigid variation, we can substitute (5.3.14) in (5.3.15) to express everything using $\mathbb{I}$ itself. This is how one gets the Arnold form in (5.3.4).

If, on the other hand, $\delta q \in \mathcal{V}_{\mathrm{INT}}$, then $\zeta := (D\mathbb{I}(q)^{-1}\delta q)\mu \in \mathfrak{g}_\mu$, so (5.3.15) gives

$$\delta^2 V_\mu(q_e) \cdot (\Delta q, \delta q) = \langle \mu, ad_\eta\zeta \rangle = -\langle ad_\zeta^*\mu, \eta \rangle$$

which vanishes since $\zeta \in \mathfrak{g}_\mu$. ■

These calculations and similar ones given below establish the block diagonal structure of $\delta^2 H_\xi(z_e)$. We shall see in Chapter 8 that discrete symmetries can produce interesting subblocking within $\delta^2 V_\mu$ on $\mathcal{V}_{\text{INT}}$. We also note that Lewis [1991] has shown how to perform these same constructions for *general* Lagrangian systems. This is important because not all systems are of the form kinetic plus potential energy. For example, gyroscopic control systems are of this sort. We shall see one example in Chapter 7 and refer to Wang and Krishnaprasad [1991] for others and some of the general theory of these systems. The hallmark of gyroscopic systems is in fact the presense of magnetic terms and we shall discuss this next in the block diagonalization context.

As far as stability is concerned, we have the following consequence of block diagonalization.

Theorem 5.3 Reduced Energy-Momentum Method *Let* $z_e = (q_e, p_e)$
*be a cotangent relative equilibrium and assume that the internal variables
are not trivial; i.e.,* $\mathcal{V}_{\text{INT}} \neq \{0\}$. *If* $\delta^2 H_\xi(z_e)$ *is definite, then it must be
positive definite. Necessary and sufficient conditions for* $\delta^2 H_\xi(z_e)$ *to be
positive definite are:*

1. The Arnold form is positive definite on $\mathcal{V}_{\text{RIG}}$ *and*

2. $\delta^2 V_\mu(q_e)$ *is positive definite on* $\mathcal{V}_{\text{INT}}$.

This follows from the fact that $\delta^2 K_\mu$ is positive definite and $\delta^2 H_\xi$ has the above block diagonal structure.

In examples, it is this form of the energy-momentum method that is often the easiest to use.

Finally in this section we should note the relation between $\delta^2 V_\xi(q_e)$ and $\delta^2 V_\mu(q_e)$. In fact, a straightforward calculation shows that

$$\delta^2 V_\mu(q_e) \cdot (\delta q, \delta q) \qquad (5.3.16)$$
$$= \delta^2 V_\xi(q_e) \cdot (\delta q, \delta q) + \langle \mathbb{I}(q_e)^{-1} \mathbf{D}\mathbb{I}(q_e) \cdot \delta q \xi, \mathbf{D}\mathbb{I}(q_e) \cdot \delta q \xi \rangle$$

and the correction term is positive. Thus, if $\delta^2 V_\xi(q_e)$ is positive definite, then so is $\delta^2 V_\mu(q_e)$, but not necessarily conversely. *Thus,* $\delta^2 V_\mu(q_e)$ *gives
sharp conditions for stability (in the sense of Theorem 5.2), while* $\delta^2 V_\xi$ *gives
only sufficient conditions.*

Using the notation $\zeta = [D\mathbb{I}^{-1}(q_e) \cdot \delta q]\mu \in \mathfrak{g}_\mu$ (see (5.3.3)), observe that the "correcting term" in (5.3.16) is given by $\langle \mathbb{I}(q_e)\zeta, \zeta \rangle = \langle\langle \zeta_Q(q_e), \zeta_Q(q_e) \rangle\rangle$.

This formula (5.3.16) is often the easiest to compute with since, as we saw with the water molecule, V_μ can be complicated compared to V_ξ.

Example *The water molecule* For the water molecule, V_{RIG} is two dimensional (as it is for any system with $G = SO(3)$). The definition gives, at a configuration $(\mathbf{r}, \mathbf{s})$, and angular momentum $\boldsymbol{\mu}$,

$$
\begin{aligned}
V_{\text{RIG}} = \; & \{(\boldsymbol{\eta} \times \mathbf{r}, \boldsymbol{\eta} \times \mathbf{s}) \mid \boldsymbol{\eta} \in \mathbb{R}^3 \quad \text{satisfies} \quad \left(\frac{m}{2}\|\mathbf{r}\|^2 + 2M\overline{m}\|\mathbf{s}\|^2\right) \boldsymbol{\eta} \cdot \boldsymbol{\mu} \\
& - \frac{m}{2}(\mathbf{r} \cdot \boldsymbol{\eta})(\mathbf{r} \cdot \boldsymbol{\mu}) + 2M\overline{m}(\mathbf{s} \cdot \boldsymbol{\mu})(\mathbf{s} \cdot \boldsymbol{\eta}) = 0\}.
\end{aligned}
$$

The condition on $\boldsymbol{\eta}$ is just the condition $\boldsymbol{\eta} \in \mathfrak{g}_\mu^\perp$. The internal space is three dimensional, the dimension of shape space. The definition gives

$$
\begin{aligned}
V_{\text{INT}} = \; & \{(\delta\mathbf{r}, \delta\mathbf{s}) \mid (\boldsymbol{\eta} \cdot \boldsymbol{\xi})(m\mathbf{r} \cdot \delta\mathbf{r} + 4M\overline{m}\mathbf{s} \cdot \delta\mathbf{s}) \\
& - \frac{m}{2}[(\delta\mathbf{r} \cdot \boldsymbol{\xi})(\mathbf{r} \cdot \boldsymbol{\eta}) + (\mathbf{r} \cdot \boldsymbol{\xi})(\delta\mathbf{r} \cdot \boldsymbol{\eta})] \\
& - 2M\overline{m}[(\delta\mathbf{s} \cdot \boldsymbol{\xi})(\mathbf{s} \cdot \boldsymbol{\eta}) + (\mathbf{s} \cdot \boldsymbol{\xi})(\delta\mathbf{s} \cdot \boldsymbol{\eta})] = 0 \\
& \text{for all} \quad \boldsymbol{\xi}, \boldsymbol{\eta} \in \mathfrak{g}_\mu^\perp\}. \; \blacklozenge
\end{aligned}
$$

5.4 The Normal Form for the Symplectic Structure

One of the most interesting aspects of block diagonalization is that the rigid-internal splitting introduced in the last section also brings the symplectic structure into normal form. We already gave the general structure of this and here we provide a few more details. We emphasise once more that this implies that the equations of motion are also put into normal form and this is useful for studying eigenvalue movement for purposes of bifurcation theory. For example, for *abelian* groups, the linearized equations of motion take the *gyroscopic form*:

$$
M\ddot{q} + S\dot{q} + \Lambda q = 0
$$

where M is a positive definite symmetric matrix (the mass matrix), Λ is symmetric (the potential term) and S is skew (the gyroscopic, or magnetic term). This second order form is particularly useful for finding eigenvalues of the linearized equations (see, for example, Bloch, Krishnaprasad, Marsden and Ratiu [1991]).

To make the normal form of the symplectic structure explicit, we shall need a preliminary result.

Lemma 5.2 *Let* $\Delta q = \eta_Q(q_e) \in \mathcal{V}_{\text{RIG}}$ *and* $\Delta z = T\alpha_\mu \cdot \Delta q \in \mathcal{S}_{\text{RIG}}$. *Then*

$$\Delta z = \text{vert}\,[\mathbb{F}L(\zeta_Q(q_e))] - T^*\eta_Q(q_e) \cdot p_e \qquad (5.4.1)$$

where $\zeta = \mathbb{I}(q_e)^{-1}ad_\eta^*\mu$ *and* vert *denotes the vertical lift.*

Proof We shall give a coordinate proof and leave it to the reader to supply an intrinsic one. In coordinates, α_μ is given by (3.3.8) as $(\alpha_\mu)_i = g_{ij}A^j{}_b\mu_a\mathbb{I}^{ab}$. Differentiating,

$$[T\alpha_\mu \cdot \nu]_i = \frac{\partial g_{ij}}{\partial q^k}\nu^k A^j{}_b\mu_a\mathbb{I}^{ab} + g_{ij}\frac{\partial A^j{}_b}{\partial q^k}\nu^k\mu_a\mathbb{I}^{ab} + g_{ij}A^j{}_b\mu_a\frac{\partial \mathbb{I}^{ab}}{\partial q^k}\nu^k. \quad (5.4.2)$$

The fact that the action consists of *isometries* gives an identity allowing us to eliminate derivatives of g_{ij}:

$$\frac{\partial g_{ij}}{\partial q^k}A^k{}_a\eta^a + g_{kj}\frac{\partial A^k{}_a}{\partial q^i}\eta^a + g_{ik}\frac{\partial A^k{}_a}{\partial q^j}\eta^a = 0. \qquad (5.4.3)$$

Substituting this in (5.4.2), taking $\nu^m = A^m_a\eta^a$, and rearranging terms gives

$$(\Delta z)_i = \left\{ g_{ik}\frac{\partial A^k{}_b}{\partial q^m}A^m_a - g_{ik}\frac{\partial A^k{}_a}{\partial q^m}A^m_b - g_{kj}\frac{\partial A^k{}_a}{\partial q^i}A^j{}_b \right\}\eta^a\mu_c\mathbb{I}^{bc}$$

$$+ g_{ij}A^j{}_b A^k{}_c\frac{\partial \mathbb{I}^{ab}}{\partial q^k}\mu_a\eta^c. \qquad (5.4.4)$$

For group actions, one has the general identity $[\eta_Q, \zeta_Q] = -[\eta, \zeta]_Q$ which gives

$$\frac{\partial A^k{}_a}{\partial q^m}A^m_b - \frac{\partial A^k{}_b}{\partial q^m}A^m_a = A^k{}_c C^c_{ab}. \qquad (5.4.5)$$

This allows one to simplify the first two terms in (5.4.4) giving

$$(\Delta z)_i = -g_{ik}A^k{}_d C^d_{ab}\eta^a\mu_c\mathbb{I}^{bc} - g_{kj}\frac{\partial A^k{}_a}{\partial q^i}A^j{}_b\eta^a\mu_c\mathbb{I}^{bc} + g_{ij}A^j{}_b A^k{}_c\frac{\partial \mathbb{I}^{ab}}{\partial q^k}\mu_a\eta^c. \quad (5.4.6)$$

Next, employ the identity (5.3.14):

$$\frac{\partial \mathbb{I}^{ab}}{\partial q^k}A^k{}_c\eta^c = C^a_{ed}\eta^e\mathbb{I}^{db} + \mathbb{I}^{ae}C^b_{de}\eta^d. \qquad (5.4.7)$$

Substituting (5.4.7) into (5.4.6), one pair of terms cancels, leaving

$$(\Delta z)_i = g_{ik}A^k{}_d C^c_{ab}\eta^a\mu_c\mathbb{I}^{bd} - g_{kj}\frac{\partial A^k{}_a}{\partial q^i}A^j{}_b\eta^a\mu_c\mathbb{I}^{bc} \qquad (5.4.8)$$

which is exactly (5.4.1) since at a relative equilibrium, $p = \alpha_\mu(q)$; *i.e.*,
$p_k = g_{kj} A^j{}_b \mathbb{I}^{bc} \mu_c$. ∎

Now we can start computing the items in the symplectic form.

Lemma 5.3 *For any $\delta z \in T_{z_e} P$,*

$$\Omega(z_e)(\Delta z, \delta z) = \langle [\mathbf{DJ}(z_e) \cdot \delta z], \eta \rangle - \langle\langle \zeta_Q(q_e), \delta q \rangle\rangle. \qquad (5.4.9)$$

Proof Again, a coordinate calculation is convenient. The symplectic form on $(\Delta p, \Delta q), (\delta p, \delta q)$ is

$$\Omega(z_e)(\Delta z, \delta z) = (\delta p)_i (\Delta q)^i - (\Delta p)_i (\delta q)^i = \delta p_i A^i{}_a \eta^a - (\Delta p)_i (\delta q^i).$$

Substituting from (5.4.8), and using

$$\langle (\mathbf{DJ} \cdot \delta z), \eta \rangle = \delta p_i A^i{}_a \eta^a + p_i \frac{\partial A^i{}_a}{\partial q^k} \delta q^k \eta^a$$

gives (5.4.9). ∎

If $\delta z \in \mathcal{S}_{\text{INT}}$, then it lies in ker $\mathbf{DJ}$, so we get the *internal-rigid interaction terms*:

$$\Omega(z_e)(\Delta z, \delta z) = -\langle\langle \zeta_Q(q_e), \delta q \rangle\rangle. \qquad (5.4.10)$$

Since these involve only δq and not δp, there is a zero in the last slot in the first row of Ω.

Lemma 5.4 *The rigid-rigid terms in Ω are*

$$\Omega(z_e)(\Delta_1 z, \Delta_2 z) = -\langle \mu, [\eta_1, \eta_2] \rangle, \qquad (5.4.11)$$

which is the coadjoint orbit symplectic structure.

Proof By Lemma 5.3, with $\zeta_1 = \mathbb{I}^{-1} ad^*_{\eta_1} \mu$,

$$\begin{aligned}
\Omega(z_e)(\Delta_1 z, \Delta_2 z) &= -\langle\langle \zeta_{1Q}(q_e), \eta_{2Q}(q_e) \rangle\rangle \\
&= -\langle ad^*_{\eta_1} \mu, \eta_2 \rangle = -\langle \mu, ad_{\eta_1} \eta_2 \rangle. \quad ∎
\end{aligned}$$

Next, we turn to the magnetic terms:

Lemma 5.5 *Let $\delta_1 z = T\alpha_\mu \cdot \delta_1 q$ and $\delta_2 z = T\alpha_\mu \cdot \delta_2 q \in \mathcal{W}_{\text{INT}}$, where $\delta_1 q, \delta_2 q \in \mathcal{V}_{\text{INT}}$. Then*

$$\Omega(z_e)(\delta_1 z, \delta_2 z) = -\mathbf{d}\alpha_\mu(\delta_1 q, \delta_2 q). \tag{5.4.12}$$

Proof Regarding α_μ as a map of Q to T^*Q, and recalling that $z_e = \alpha_\mu(q_e)$, we recognize the left hand side of (5.4.12) as the pull back of Ω by α_μ (and then restricted to $\mathcal{V}_{\text{INT}}$). However, as we saw already in Chapter 2, the canonical one-form is characterized by $\beta^*\Theta = \beta$, so $\beta^*\Omega = -\mathbf{d}\beta$ for any one-form β. Therefore, $\Omega(z_e)(\delta_1 z, \delta_2 z) = -\mathbf{d}\alpha_\mu(q_e)(\delta_1 q, \delta_2 q)$. ∎

If we define the one form α_ξ by $\alpha_\xi(q) = \mathbb{F}L(\xi_Q(q))$, then the definition of $\mathcal{V}_{\text{INT}}$ shows that on this space $\mathbf{d}\alpha_\mu = \mathbf{d}\alpha_\xi$. This is a useful remark since $\mathbf{d}\alpha_\xi$ is somewhat easier to compute in examples.

If we had made the "naive" choice of $\mathcal{V}_{\text{INT}}$ as the orthogonal complement of the G-orbit, then we could also replace $\mathbf{d}\alpha_\mu$ by $\langle \mu, \text{curv}\alpha \rangle$. However, with our choice of $\mathcal{V}_{\text{INT}}$, one must be careful of the distinction.

We leave it for the reader to check that the rest of the $\mathcal{W}_{\text{INT}}, \mathcal{W}^*_{\text{INT}}$ block is as stated.

5.5   Stability of Relative Equilibria for the Double Spherical Pendulum

We now give some of the results for the stability of the branches of the double spherical pendulum that we found in the last chapter. We refer the reader to Marsden and Scheurle [1992] for additional details. Even though the symmetry group of this example is abelian, and so there is no rigid body block, the calculations are by no means trivial. We shall leave the simple pendulum to the reader, in which case all of the relative equilibria are stable, except for the straight upright solution, which is unstable.

The water molecule is a nice illustration of the general structure of the method. We shall not work this example out here, however, as it is quite complicated, as we have mentioned. However, we shall come back to it in Chapter 8 and indicate some more general structure that can be obtained on grounds of discrete symmetry alone. For other informative examples, the reader can consult Lewis and Simo [1990], Simo, Lewis and Marsden [1991], Simo, Posberg and Marsden [1990, 1991], and Zombro and Holmes [1991].

To carry out the stability analysis for relative equilibria of the double spherical pendulum, one must compute $\delta^2 V_\mu$ on the subspace orthogonal to the G_μ-orbit. To do this, is is useful to introduce coordinates adapted to the problem and to work in Lagrangian representation. Specifically, let $\mathbf{q}_1^\perp$ and $\mathbf{q}_2^\perp$ be given polar coordinates (r_1, θ_1) and (r_2, θ_2) respectively. Then $\varphi = \theta_2 - \theta_1$ represents an S^1-invariant coordinate, the angle between the two vertical planes formed by the pendula. In these terms, one computes from our earlier expressions that the angular momentum is

$$
\begin{aligned}
J &= (m_1 + m_2)r_1^2\dot{\theta}_1 + m_2 r_2^2 \dot{\theta}_2 \\
&= + m_2 r_1 r_2 (\dot{\theta}_1 + \dot{\theta}_2)\cos\varphi + m_2 (r_1 \dot{r}_2 - r_2 \dot{r}_1)\sin\varphi
\end{aligned}
\tag{5.5.1}
$$

and the Lagrangian is

$$
\begin{aligned}
L &= \frac{1}{2}m_1(\dot{r}_1^2 + r_1^2\dot{\theta}_1^2) + \frac{1}{2}m_2\left\{\dot{r}_1^2 + r_1^2\dot{\theta}_1^2 + \dot{r}_2^2 + r_2^2\dot{\theta}_2^2\right. \\
&\quad \left. + 2(\dot{r}_1\dot{r}_2 + r_1 r_2 \dot{\theta}_1\dot{\theta}_2)\cos\varphi + 2(r_1\dot{r}_2\dot{\theta}_1 - r_2\dot{r}_1\dot{\theta}_2)\sin\varphi\right\} \\
&\quad + \frac{1}{2}m_1\frac{r_1^2\dot{r}_1^2}{l_1^2 - r_1^2} + \frac{1}{2}m_2\left(\frac{r_1\dot{r}_1}{\sqrt{l_1^2 - r_1^2}} + \frac{r_2\dot{r}_2}{\sqrt{l_2^2 - r_2^2}}\right)^2 \\
&\quad - m_1 g\sqrt{l_1^2 - r_1^2} - m_2 g\left(\sqrt{l_1^2 - r_1^2} + \sqrt{l_2^2 - r_2^2}\right).
\end{aligned}
\tag{5.5.2}
$$

One also has, from (3.5.17),

$$
\begin{aligned}
V_\mu &= -m_1 g\sqrt{l_1^2 - r_1^2} - m_2 g\left(\sqrt{l_1^2 - r_1^2} + \sqrt{l_2^2 - r_2^2}\right) \\
&\quad + \frac{1}{2}\frac{\mu^2}{m_1 r_1^2 + m_2(r_1^2 + r_2^2 + 2r_1 r_2\cos\varphi)}.
\end{aligned}
\tag{5.5.3}
$$

Notice that V_μ depends on the angles θ_1 and θ_2 only through $\varphi = \theta_2 - \theta_1$, as it should by S^1-invariance. Next one calculates the second variation at one of the relative equilibria found in §4.3. If we calculate it as a 3×3 matrix in the variables r_1, r_2, φ, then one checks that we will automatically be in a space orthogonal to the G_μ-orbits. One finds, after some computation, that

$$
\delta^2 V_\mu = \begin{bmatrix} a & b & 0 \\ b & d & 0 \\ 0 & 0 & e \end{bmatrix}
\tag{5.5.4}
$$

where

$$
\begin{aligned}
a &= \frac{\mu^2(3(m+\alpha)^2 - \alpha^2(m-1))}{\lambda^4 l_1^2 m_2(m + \alpha^2 + 2\alpha)^3} + \frac{gm_2 m}{l_1(1 + \lambda^2)^{3/2}} \\
b &= (\operatorname{sign}\alpha)\frac{\mu^2}{\lambda^4 l_1^4 m_2}\frac{3(m + \alpha^2 + 2\alpha) + 4\alpha(m-1)}{(m + \alpha^2 + 2\alpha)^3}
\end{aligned}
$$

$$d = \frac{\mu^2}{\lambda^4 l_1^4 m_2} \frac{3(\alpha+1)^2 + 1 - m}{(m + \alpha^2 + 2\alpha)^3} + \frac{m_2 g}{l_1} \frac{r^2}{(r^2 - \lambda^2 \alpha^2)^{3/2}}$$

$$e = \frac{\mu^2}{\lambda^2 l_1^2 m_2} \frac{\alpha}{(m + \alpha^2 + 2\alpha)^2}.$$

Notice the zeros in (5.5.4); they are in fact a result of discrete symmetry, as we shall see in Chapter 8. Without the help of these zeros (for example, if the calculation is done in arbitrary coordinates), the expression for $\delta^2 V_\mu$ might be intractible.

Based on this calculation one finds:

Proposition 5.4 *The signature of $\delta^2 V_\mu$ along the "straight out" branch of the double spherical pendulum (with $\alpha > 0$) is $(+,+,+)$ and so is stable. The signature along the "cowboy branch" with $\alpha < 0$ and emanating from the straight down state $(\lambda = 0)$ is $(-,-,+)$ and along the remaining branches is $(-,+,+)$.*

The stability along the cowboy branch requires further analysis that we shall outline in Chapter 10. The remaining branches are linearly unstable since the index is odd (this is for reasons we shall go into in Chapter 10).

To get this instability and bifurcation information, one needs to linearize the reduced equations and compute the corresponding eigenvalues. There are (at least) three methodologies that can be used for computing the reduced linearized equations:

i   Compute the Euler-Lagrange equations from (5.5.2), drop them to $\mathbf{J}^{-1}(\mu)/G_\mu$ and linearize the resulting equations.

ii   Read off the linearized reduced equations from the block diagonal form of $\delta^2 H_\xi$ and the symplectic structure.

iii   Perform Lagrangian reduction as in §3.6 to obtain the Lagrangian structure of the reduced system and linearize it at a relative equilibrium.

For the double spherical pendulum, perhaps the first method is the quickest to get the answer, but of course the other methods provide insight and information about the structure of the system obtained.

The linearized system obtained has the following standard form expected for abelian reduction:

$$M\ddot{q} + S\dot{q} + \Lambda q = 0. \qquad (5.5.5)$$

In our case $q = (r_1, r_2, \varphi)$ and Λ is the matrix (5.5.4) given above. The mass matrix M is

$$M = \begin{bmatrix} m_{11} & m_{12} & 0 \\ m_{12} & m_{22} & 0 \\ 0 & 0 & m_{33} \end{bmatrix}$$

where

$$m_{11} = \frac{m_1 + m_2}{1 - \lambda^2}, \quad m_{12} = (\text{sign } \alpha) m_2 \left(1 + \frac{\alpha \lambda^2}{\sqrt{1 - \lambda^2} \sqrt{r^2 - \alpha^2 \lambda^2}} \right)$$

$$m_{22} = m_2 \frac{r^2}{r^2 - \lambda^2 \alpha^2}, \quad m_{33} = m_2 l_1^2 \lambda^2 (m - 1) \frac{\alpha^2}{m + \alpha^2 + 2\alpha}$$

and the gyroscopic matrix S (the magnetic term) is

$$S = \begin{bmatrix} 0 & 0 & s_{13} \\ 0 & 0 & s_{23} \\ -s_{13} & -s_{23} & 0 \end{bmatrix}$$

where

$$s_{13} = \frac{\mu}{\lambda l_1} \frac{2\alpha^2 (m - 1)}{(m + \alpha^2 + 2\alpha)^2} \quad \text{and}$$

$$s_{23} = -(\text{sign } \alpha) \frac{\mu}{\lambda l_1} \frac{2\alpha(m - 1)}{(m + \alpha^2 + 2\alpha)^2}.$$

We will pick up this discussion again in Chapter 10.

Chapter 6

Geometric Phases

In this chapter we give the basic ideas for geometric phases in terms of reconstruction, prove Montgomery's formula for rigid body phases, and give some other basic examples. We refer to Marsden, Montgomery and Ratiu [1990] for more information.

6.1   A Simple Example

In Chapter 1 we discussed the idea of phases and gave several examples in general terms. What follows is a specific but very simple example to illustrate the important role played by the conserved quantity (in this case the angular momentum).

Consider two planar rigid bodies joined by a pin joint *at their centers of mass*. Let I_1 and I_2 be their moments of inertia and θ_1 and θ_2 be the angle they make with a fixed *inertial* direction, as in Figure 6.1.1.

Conservation of angular momentum states that $I_1\dot{\theta}_1 + I_2\dot{\theta}_2 = \mu = $ constant in time, where the overdot means time derivative. Recall from Chapter 3 that the *shape space* Q/G of a system is the space whose points give the shape of the system. In this case, shape space is the circle S^1 parametrized by the *hinge angle* $\psi = \theta_2 - \theta_1$. We can parametrize the configuration space of the system by θ_1 and θ_2 or by $\theta = \theta_1$ and ψ. Conservation of angular momentum reads

$$I_1\dot{\theta} + I_2(\dot{\theta} + \dot{\psi}) = \mu; \quad \text{that is,} \quad d\theta + \frac{I_2}{I_1 + I_2}d\psi = \frac{\mu}{I_1 + I_2}dt. \quad (6.1.1)$$

The left hand side of (6.1.1) is the *mechanical connection* discussed in Chapter 3. Suppose that body #2 goes through one full revolution so that ψ

115

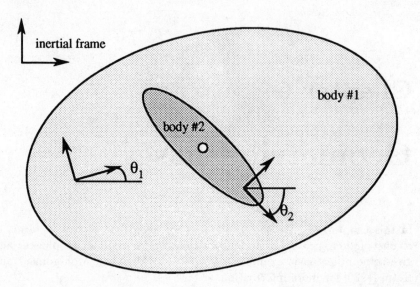

Figure 6.1.1: Two rigid bodies coupled at their centers of mass.

increases from 0 to 2π. Suppose, moreover, that the total angular momentum is zero: $\mu = 0$. From (6.1.1) we see that

$$\Delta\theta = -\frac{I_2}{I_1 + I_2} \int_0^{2\pi} d\psi = -\left(\frac{I_2}{I_1 + I_2}\right) 2\pi. \qquad (6.1.2)$$

This is the amount by which body #1 rotates, each time body #2 goes around once. This result is independent of the detailed dynamics and only depends on the fact that angular momentum is conserved and that body #2 goes around once. In particular, we get the same answer even if there is a "hinge potential" hindering the motion or if there is a control present in the joint. Also note that if we want to rotate body #1 by $-2\pi k I_2/(I_1 + I_2)$ radians, where k is an integer, all one needs to do is spin body #2 around k times, then stop it. By conservation of angular momentum, body #1 will stay in that orientation after stopping body #2.

In particular, if we think of body #1 as a spacecraft and body #2 as an internal rotor, this shows that by manipulating the rotor, we have *control* over the attitude (orientation) of the spacecraft.

Here is a geometric interpretation of this calculation. Define the one form

$$A = d\theta + \frac{I_2}{I_1 + I_2}d\psi. \tag{6.1.3}$$

This is a flat connection for the trivial principal S^1-bundle $\pi : S^1 \times S^1 \to S^1$ given by $\pi(\theta, \psi) = \psi$. Formula (6.1.2) is the holonomy of this connection, when we traverse the base circle, $0 \le \psi \le 2\pi$.

Another interesting context in which geometric phases comes up is the phase shift that occurs in interacting solitons. In fact, Alber and Marsden [1992] have shown how this is a geometric phase in the sense described in this chapter. In the introduction we already discussed a variety of other examples.

6.2 Reconstruction

This section presents a reconstruction method for the dynamics of a given Hamiltonian system from that of the reduced system. Let P be a symplectic manifold on which a Lie group acts in a Hamiltonian manner and has a momentum map $\mathbf{J} : P \to \mathfrak{g}^*$. Assume that an integral curve $c_\mu(t)$ of the reduced Hamiltonian vector field X_{H_μ} on the reduced space P_μ is known. For $z_0 \in \mathbf{J}^{-1}(\mu)$, we search for the corresponding integral curve $c(t) = F_t(z_0)$ of X_H such that $\pi_\mu(c(t)) = c_\mu(t)$, where $\pi_\mu : \mathbf{J}^{-1}(\mu) \to P_\mu$ is the projection.

To do this, choose a smooth curve $d(t)$ in $\mathbf{J}^{-1}(\mu)$ such that $d(0) = z_0$ and $\pi_\mu(d(t)) = c_\mu(t)$. Write $c(t) = \Phi_{g(t)}(d(t))$ for some curve $g(t)$ in G_μ to be determined, where the group action is denoted $g \cdot z = \Phi_g(z)$. First note that

$$
\begin{aligned}
X_H(c(t)) &= c'(t) \\
&= T_{d(t)}\Phi_{g(t)}(d'(t)) \\
&\quad + T_{d(t)}\Phi_{g(t)} \cdot \left(T_{g(t)}L_{g(t)^{-1}}(g'(t))\right)_P (d(t)). \tag{6.2.1}
\end{aligned}
$$

Since $\Phi_g^* X_H = X_{\Phi_g^* H} = X_H$, (6.2.1) gives

$$
\begin{aligned}
d'(t) &+ \left(T_{g(t)}L_{g(t)^{-1}}(g')(t)\right)_P (d(t)) \\
&= T\Phi_{g(t)^{-1}} X_H(\Phi_{g(t)}(d(t))) \\
&= (\Phi_{g(t)}^* X_H)(d(t)) \\
&= X_H(d(t)). \tag{6.2.2}
\end{aligned}
$$

This is an equation for $g(t)$ written in terms of $d(t)$ only. We solve it in two steps:

Step 1 *Find* $\xi(t) \in \mathfrak{g}_\mu$ *such that*

$$\xi(t)_P(d(t)) = X_H(d(t)) - d'(t). \tag{6.2.3}$$

Step 2 *With $\xi(t)$ determined, solve the following non-autonomous ordinary differential equation on G_μ:*

$$g'(t) = T_e L_{g(t)}(\xi(t)), \quad \text{with} \quad g(0) = e. \tag{6.2.4}$$

Step 1 is typically of an algebraic nature; in coordinates, for matrix Lie groups, (6.2.3) is a matrix equation. We show later how $\xi(t)$ can be explicitly computed if a connection is given on $\mathbf{J}^{-1}(\mu) \to P_\mu$. With $g(t)$ determined, the desired integral curve $c(t)$ is given by $c(t) = \Phi_{g(t)}(d(t))$. A similar construction works on P/G, even if the G-action does not admit a momentum map.

Step 2 can be carried out *explicitly* when G is abelian. Here the connected component of the identity of G is a cylinder $\mathbb{R}^p \times \mathbb{T}^{k-p}$ and the exponential map $\exp(\xi_1, \ldots, \xi_k) = (\xi_1, \ldots, \xi_p, \xi_{p+1}(\mathrm{mod}2\pi), \ldots, \xi_k(\mathrm{mod}2\pi))$ is onto, so we can write $g(t) = \exp \eta(t)$, where $\eta(0) = 0$. Therefore $\xi(t) = T_{g(t)} L_{g(t)^{-1}}(g'(t)) = \eta'(t)$ since η' and η commute, *i.e.*, $\eta(t) = \int_0^t \xi(s) ds$. Thus the solution of (6.2.4) is

$$g(t) = \exp \left(\int_0^t \xi(s) ds \right). \tag{6.2.5}$$

This reconstruction method depends on the choice of $d(t)$. With additional structure, $d(t)$ can be chosen in a *natural geometric* way. What is needed is a way of lifting curves on the base of a principal bundle to curves in the total space, which can be done using connections. One can object at this point, noting that reconstruction involves integrating *one* ordinary differential equation, whereas introducing a connection will involve integration of *two* ordinary differential equations, one for the horizontal lift and one for constructing the solution of (6.2.4) from it. However, for the determination of phases, there are some situations in which the phase can be computed without actually solving either equation, so one actually solves *no* differential equations; a specific case is the rigid body.

Suppose that $\pi_\mu : \mathbf{J}^{-1}(\mu) \to P_\mu$ is a principal G_μ-bundle with a *connection A*. This means that A is a $\mathfrak{g}_\mu$-valued one-form on $\mathbf{J}^{-1}(\mu) \subset P$ satisfying

  i $A_p \cdot \xi_P(p) = \xi$ for $\xi \in \mathfrak{g}_\mu$
  ii $L_g^* A = Ad_g \circ A$ for $g \in G_\mu$.

Let $d(t)$ be the **horizontal lift** of c_μ through z_0, *i.e.*, $d(t)$ satisfies $A(d(t)) \cdot d'(t) = 0, \pi_\mu \circ d = c_\mu$, and $d(0) = z_0$. We summarize:

Theorem 6.1 Reconstruction *Suppose* $\pi_\mu : \mathbf{J}^{-1}(\mu) \to P_\mu$ *is a principal G_μ-bundle with a connection A. Let c_μ be an integral curve of the reduced dynamical system on P_μ. Then the corresponding curve c through a point $z_0 \in \pi_\mu^{-1}(c_\mu(0))$ of the system on P is determined as follows:*

  i   *Horizontally lift c_μ to form the curve d in $\mathbf{J}^{-1}(\mu)$ through z_0.*

  ii   *Let $\xi(t) = A(d(t)) \cdot X_H(d(t))$, so that $\xi(t)$ is a curve in $\mathfrak{g}_\mu$.*

 iii *Solve the equation $\dot{g}(t) = g(t) \cdot \xi(t)$ with $g(0) = e$.*

Then $c(t) = g(t) \cdot d(t)$ is the integral curve of the system on P with initial condition z_0.

Suppose c_μ is a *closed* curve with period T; thus, both c and d reintersect the same fiber. Write

$$d(T) = \hat{g} \cdot d(0) \quad \text{and} \quad c(T) = h \cdot c(0)$$

for $\hat{g}, h \in G_\mu$. Note that

$$h = g(T)\hat{g}. \tag{6.2.6}$$

The Lie group element $\hat{g}$ (or the Lie algebra element $\log g$) is called the **geometric phase**. It is the holonomy of the path c_μ with respect to the connection A and has the important property of being parametrization independent. The Lie group element $g(T)$ (or $\log g(T)$) is called the **dynamic phase**.

For compact or semi-simple G, the group G_μ is generically abelian. The computation of $g(T)$ and $\hat{g}$ are then relatively easy, as was indicated above.

6.3   Cotangent Bundle Phases — a Special Case

We now discuss the case in which $P = T^*Q$, and G acts on Q and therefore on P by cotangent lift. In this case the momentum map is given by the formula

$$\mathbf{J}(\alpha_q) \cdot \xi = \alpha_q \cdot \xi_Q(q) = \xi_{T^*Q}(\alpha_q) \, \lrcorner\, \theta(\alpha_q),$$

where $\xi \in \mathfrak{g}_\mu, \alpha_q \in T_q^*Q, \theta = p_i dq^i$ is the canonical one-form and $\lrcorner$ is the interior product.

Assume G_μ is the circle group or the line. Pick a generator $\zeta \in \mathfrak{g}_\mu, \zeta \neq 0$. For instance, one can choose the shortest ζ such that $\exp(2\pi\zeta) = 1$. Identify $\mathfrak{g}_\mu$ with the real line via $\omega \mapsto \omega\zeta$. Then a connection one-form is a standard one-form on $\mathbf{J}^{-1}(\mu)$.

Proposition 6.1 *Suppose $G_\mu \cong S^1$ or $\mathbb{R}$. Identify $\mathfrak{g}_\mu$ with $\mathbb{R}$ via a choice of generator ζ. Let θ_μ denote the pull-back of the canonical one-form to $\mathbf{J}^{-1}(\mu)$. Then*

$$A = \frac{1}{\langle \mu, \zeta \rangle} \theta_\mu \otimes \zeta$$

is a connection one-form on $\mathbf{J}^{-1}(\mu) \to P_\mu$. Its curvature as a two-form on the base P_μ is

$$\Omega = -\frac{1}{\langle \mu, \zeta \rangle} \omega_\mu,$$

where ω_μ is the reduced symplectic form on P_μ.

Proof Since G acts by cotangent lift, it preserves θ, and so θ_μ is preserved by G_μ and therefore A is G_μ-invariant. Also, $A \cdot \zeta_P = [\zeta_P \lrcorner \theta / \langle \mu, \zeta \rangle] \zeta = [\langle \mathbf{J}, \zeta \rangle / \langle \mu, \zeta \rangle] \zeta = \zeta$. This verifies that A is a connection. The calculation of its curvature is straightforward. ∎

Remarks
1 The result of Proposition 6.1 holds for any *exact* symplectic manifold. We shall use this for the rigid body.
2 In the next section we shall show how to construct a connection on $\mathbf{J}^{-1}(\mu) \to P_\mu$ in general. For the rigid body it is easy to check that *these two constructions coincide!* In general, they coincide when $G = Q$ and $G_\mu = S^1$.
3 Choose P to be a complex Hilbert space with symplectic form $\Omega(\varphi, \psi) = -\mathrm{Im}\langle \varphi, \psi \rangle$ and S^1 action given by $e^{i\theta}\psi$. The corresponding momentum map is $J(\varphi) = -\|\varphi\|^2/2$. Write $\Omega = -d\Theta$ where $\Theta(\varphi) \cdot \varphi = \frac{1}{2}\mathrm{Im}\langle \varphi, \psi \rangle$. Now we identify the reduced space at level $-1/2$ to get projective Hilbert space. Applying Proposition 6.1, *we get the basic phase result for quantum mechanics* due to Aharonov and Anandan [1987].

> The holonomy of a loop in projective complex Hilbert space is the exponential of twice the symplectic area of any two dimensional submanifold whose boundary is the given loop.

6.4   Cotangent Bundles — General Case

If G_μ is not abelian, the formula for A given above does not satisfy the second axiom of a connection. However, if the bundle $Q \to Q/G_\mu$ has a connection, we will show below how this induces a connection on $\mathbf{J}^{-1}(\mu) \to (T^*Q)_\mu$. To do this, we can use the cotangent bundle reduction theorem.

Recall that the mechanical connection provides a connection on $Q \to Q/G$ and also on $Q \to Q/G_\mu$. Denote the latter connection by γ.

Denote by $\mathbf{J}_\mu : T^*Q \to \mathfrak{g}_\mu^*$ the induced momentum map, *i.e.*, $\mathbf{J}_\mu(\alpha_q) = \mathbf{J}(\alpha_q)|\mathfrak{g}_\mu$. From the cotangent bundle reduction theorem, it follows that the diagram in Figure 6.4.1 commutes.

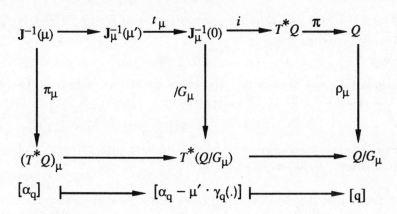

Figure 6.4.1: Notation for cotangent bundles.

In this figure, $t_\mu(\alpha_q) = \alpha_q - \mu' \cdot \gamma_q(\cdot)$ is fiber translation by the μ'-component of the connection form and where $[\alpha_q - \mu' \cdot \gamma_q(\cdot)]$ means the element of $T^*(Q/G_\mu)$ determined by $\alpha_q - \mu' \cdot \gamma_q(\cdot)$ and $\mu' = \mu \mid \mathfrak{g}_\mu$. Call the composition of the two maps on the bottom of this diagram

$$\sigma : [\alpha_q] \in (T^*Q)_\mu \mapsto [q] \in Q/G_\mu.$$

We induce a connection on $\mathbf{J}^{-1}(\mu) \to (T^*Q)_\mu$ by being consistent with this diagram.

Proposition 6.2 *The connection one-form γ induces a connection one-form $\tilde{\gamma}$ on $\mathbf{J}^{-1}(\mu)$ by pull-back: $\tilde{\gamma} = (\pi \circ t_\mu)^*\gamma$, i.e.,*

$$\tilde{\gamma}(\alpha_q) \cdot U_{\alpha_q} = \gamma(q) \cdot T_{\alpha_q}\pi(U_{\alpha_q}), \quad \text{for} \quad \alpha_q \in T_q^*Q, U_{\alpha_q} \in T_{\alpha_q}(T^*Q).$$

Similarly curv $(\tilde{\gamma}) = (\pi \circ t_\mu)^*$curv$(\gamma)$ *and in particular the μ'-component of the curvature of this connection is the pull-back of $\mu' \cdot$ curv(γ).*

The proof is a direct verification.

Proposition 6.3 *Assume that $\rho_\mu : Q \to Q/G_\mu$ is a principal G_μ-bundle with a connection γ. If H is a G-invariant Hamiltonian on T^*Q inducing* *the Hamiltonian H_μ on $(T^*Q)_\mu$ and $c_\mu(t)$ is an integral curve of X_{H_μ} denote by $d(t)$ a horizontal lift of $c_\mu(t)$ in $\mathbf{J}^{-1}(\mu)$ relative to the natural connection of Proposition 6.2 and let $q(t) = \pi(d(t))$ be the base integral curve of $c(t)$. Then $\xi(t)$ of step ii in Theorem 6.1 is given by*

$$\xi(t) = \gamma(q(t)) \cdot \mathbb{F}H(d(t)),$$

where $\mathbb{F}H : T^*Q \to TQ$ is the fiber derivative of H, i.e., $\mathbb{F}H(\alpha_q) \cdot \beta_q = \frac{d}{dt} H(\alpha_q + t\beta_q)\big|_{t=0}$.

Proof $\tilde{\gamma} \cdot X_H = \gamma \cdot T\pi \cdot X_H = \gamma \cdot \mathbb{F}H$. ∎

Proposition 6.4 *With γ the mechanical connection, step ii in Theorem 6.1 is equivalent to ii' $\xi(t) \in \mathfrak{g}_\mu$ is given by*

$$\xi(t) = \gamma(q(t)) \cdot d(t)^\sharp, \quad \text{where} \quad d(t) \in T^*_{q(t)}Q.$$

Proof Apply Proposition 6.2 and use the fact that $\mathbb{F}H(\alpha_q) = a_q^\sharp$. (In coordinates, $\partial H/\partial p_\mu = g^{\mu\nu}p_\nu$.) ∎

To see that the connection of Proposition 6.2 coincides with the one in §6.3 for the rigid body, use the fact that $Q = G$ and α_μ must lie in $\mathbf{J}^{-1}(\mu)$, so α_μ is the right invariant one form equalling μ at $g = e$ and that $G_\mu = S^1$.

6.5 Rigid Body Phases

We now derive a formula of Montgomery [1991] for the rigid body phase (see also Marsden, Montgomery and Ratiu [1990]).

As we have seen, the motion of a rigid body is a geodesic with respect to a left-invariant Riemannian metric (the inertia tensor) on $SO(3)$. The corresponding phase space is $P = T^*SO(3)$ and the momentum map $\mathbf{J} : P \to \mathbb{R}^3$ for the *left* $SO(3)$ action is *right* translation to the identity. We identify $so(3)^*$ with $so(3)$ via the Killing form and identify $\mathbb{R}^3$ with $so(3)$ via the map $v \mapsto \hat{v}$ where $\hat{v}(w) = v \times w$, $\times$ being the standard cross product. Points in $so(3)^*$ are regarded as the left reduction of $T^*SO(3)$ by $SO(3)$ and are the angular momenta as seen from a *body-fixed* frame. The reduced spaces $\mathbf{J}^{-1}(\mu)/G_\mu$ are identified with spheres in $\mathbb{R}^3$ of Euclidean radius $\|\mu\|$, with their symplectic form $\omega_\mu = -dS/\|\mu\|$ where dS is the standard area form on a sphere of radius $\|\mu\|$ and where G_μ consists of rotations about the μ-axis. The trajectories of the reduced dynamics are obtained by intersecting a family of homothetic ellipsoids (the energy ellipsoids) with the angular momentum spheres. In particular, all but at most four of the reduced trajectories are periodic. These four exceptional trajectories are the homoclinic trajectories.

Suppose a reduced trajectory $\Pi(t)$ is given on P_μ, with period T. *After time T, by how much has the rigid body rotated in space?*

The spatial angular momentum is $\pi = \mu = g\Pi$, which is the conserved value of $\mathbf{J}$. Here $g \in SO(3)$ is the attitude of the rigid body and Π is the body angular momentum. If $\Pi(0) = \Pi(T)$ then

$$\mu = g(0)\Pi(0) = g(T)\Pi(T)$$

and so

$$g(T)^{-1}\mu = g(0)^{-1}\mu \quad i.e., \quad g(T)g(0)^{-1}$$

is a rotation about the axis μ. We want to compute the angle of this rotation.

To answer this question, let $c(t)$ be the corresponding trajectory in $\mathbf{J}^{-1}(\mu) \subset P$. Identify $T^*SO(3)$ with $SO(3) \times \mathbb{R}^3$ by left trivialization, so $c(t)$ gets identified with $(g(t), \Pi(t))$. Since the reduced trajectory $\Pi(t)$ closes after time T, we recover the fact that $c(T) = gc(0)$ where $g = g(T)g(0)^{-1} \in G_\mu$. Thus, we can write

$$g = \exp[(\Delta\theta)\zeta] \tag{6.5.1}$$

where $\zeta = \mu/\|\mu\|$ identifies $\mathfrak{g}_\mu$ with $\mathbb{R}$ by $a\zeta \mapsto a$, for $a \in \mathbb{R}$. Let D be one of the two spherical caps on S^2 enclosed by the reduced trajectory, Λ be the corresponding *oriented* solid angle, with $|\Lambda| = (\text{area } D)/\|\mu\|^2$, and let H_μ be the energy of the reduced trajectory. All norms are taken relative to the Euclidean metric of $\mathbb{R}^3$. We shall prove below that modulo 2π, we have

$$\Delta\theta = \frac{1}{\|\mu\|}\left\{\int_D \omega_\mu + 2H_\mu T\right\} = -\Lambda + \frac{2H_\mu T}{\|\mu\|}. \tag{6.5.2}$$

(The special case of this formula for a *symmetric* free rigid body was given by Hannay [1985] and Anandan [1988].)

To prove (6.5.2), we choose the connection one-form on $\mathbf{J}^{-1}(\mu)$ to be the one from Proposition 6.1, or equivalently from Proposition 6.2:

$$A = \frac{1}{\|\mu\|}\theta_\mu, \tag{6.5.3}$$

where θ_μ is the pull back to $\mathbf{J}^{-1}(\mu)$ of the canonical one-form θ on $T^*SO(3)$. The curvature of A as a two-form on the base P_μ, the sphere of radius $\|\mu\|$ in $\mathbb{R}^3$, is given by

$$-\frac{1}{\|\mu\|}\omega_\mu = \frac{1}{\|\mu\|^2}dS. \tag{6.5.4}$$

The first terms in (6.5.2) represent the *geometric phase*, *i.e.*, the holonomy of the reduced trajectory with respect to this connection. The logarithm of the holonomy (modulo 2π) is given as minus the integral over D of the curvature, *i.e.*, it equals

$$\frac{1}{\|\mu\|} \int_D \omega_\mu = -\frac{1}{\|\mu\|^2}(\text{area } D) = -\Lambda(\text{mod } 2\pi). \qquad (6.5.5)$$

The second terms in (6.5.2) represent the *dynamic phase*. By Theorem 6.1, it is calculated in the following way. First one horizontally lifts the reduced closed trajectory $\Pi(t)$ to $J^{-1}(\mu)$ relative to the connection (6.5.3). This horizontal lift is easily seen to be (identity, $\Pi(t)$) in the left trivialization of $T^*SO(3)$ as $SO(3) \times \mathbb{R}^3$. Second, we need to compute

$$\xi(t) = (A \cdot X_H)(\Pi(t)). \qquad (6.5.6)$$

Since in coordinates

$$\theta_\mu = p_i dq^i \quad \text{and} \quad X_H = p^i \frac{\partial}{\partial q^i} + \frac{\partial}{\partial p} \quad \text{terms}$$

for $p^i = \sum_j g^{ij} p_j$, where g^{ij} is the inverse of the Riemannian metric g_{ij} on $SO(3)$, we get

$$(\theta_\mu \cdot X_H)(\Pi(t)) = p_i p^i = 2H(\text{identity}, \Pi(t)) = 2H_\mu, \qquad (6.5.7)$$

where H_μ is the value of the energy on S^2 along the integral curve $\Pi(t)$. Consequently,

$$\xi(t) = \frac{2H_\mu}{\|\mu\|}\zeta. \qquad (6.5.8)$$

Third, since $\xi(t)$ is independent of t, the solution of the equation

$$\dot{g} = g\xi = \frac{2H_\mu}{\|\mu\|}g\zeta \quad \text{is} \quad g(t) = \exp\left(\frac{2H_\mu t}{\|\mu\|}\zeta\right)$$

so that the dynamic phase equals

$$\Delta\theta_d = \frac{2H_\mu}{\|\mu\|}T(\text{mod } 2\pi). \qquad (6.5.9)$$

Formulas (6.5.5) and (6.5.9) prove (6.5.2). Note that (6.5.2) is independent of which spherical cap one chooses amongst the two bounded by $\Pi(t)$. Indeed, the solid angles on the unit sphere defined by the two caps add to 4π, which does not change formula (6.5.2).

For other examples of the use of (6.5.2) see Chapter 7 and Levi [1991]. We also note that Levi gives an interesting link between this result, the Poinsot description of rigid body motion and the Gauss-Bonnet theorem.

6.6 Moving Systems

The techniques above can be merged with those for adiabatic systems, with slowly varying parameters. We illustrate the ideas with the example of the bead in the hoop discussed in Chapter 1.

Begin with a reference configuration Q and a Riemannian manifold S. Let M be a space of embeddings of Q into S and let m_t be a curve in M. If a particle in Q is following a curve $q(t)$, and if we let the configuration space Q have a superimposed motion m_t, then the path of the particle in S is given by $m_t(q(t))$. Thus, its velocity in S is given by the time derivative:

$$T_{q(t)}m_t \cdot \dot{q}(t) + \mathcal{Z}_t(m_t(q(t))) \tag{6.6.1}$$

where $\mathcal{Z}_t$, defined by $\mathcal{Z}_t(m_t(q)) = \frac{d}{dt}m_t(q)$, is the time dependent vector field (on S with domain $m_t(Q)$) generated by the motion m_t and $T_{q(t)}m_t \cdot w$ is the derivative (tangent) of the map m_t at the point $q(t)$ in the direction w. To simplify the notation, we write

$$\mathfrak{m}_t = T_{q(t)}m_t \quad \text{and} \quad \mathsf{q}(t) = m_t(q(t)).$$

Consider a Lagrangian on TQ of the form kinetic minus potential energy. Using (6.6.1), we thus choose

$$L_{m_t}(q, v) = \frac{1}{2}\|\mathfrak{m}_t \cdot v + \mathcal{Z}_t(\mathsf{q}(t))\|^2 - V(q) - U(\mathsf{q}(t)) \tag{6.6.2}$$

where V is a given potential on Q and U is a given potential on S.

Put on Q the (possibly time dependent) metric induced by the mapping m_t. In other words, choose the metric on Q that makes m_t into an isometry for each t. In many examples of mechanical systems, such as the bead in the hoop given below, m_t is already a restriction of an isometry to a submanifold of S, so the metric on Q in this case is in fact time independent. Now we take the Legendre transform of (6.6.2), to get a Hamiltonian system on T^*Q. Taking the derivative of (6.6.2) with respect to v in the direction of w gives:

$$p \cdot w = \langle \mathfrak{m}_t \cdot v + \mathcal{Z}_t(q(t)), \mathfrak{m}_t \cdot w \rangle_{\mathsf{q}(t)} = \langle \mathfrak{m}_t \cdot v + \mathcal{Z}_t(q(t))^T, \mathfrak{m}_t \cdot w \rangle_{\mathsf{q}(t)}$$

where $p \cdot w$ means the natural pairing between the covector $p \in T^*_{q(t)}Q$ and the vector $w \in T_{q(t)}Q$, $\langle\,,\,\rangle_{\mathsf{q}(t)}$ denotes the metric inner product on the space S at the point $\mathsf{q}(t)$ and ^{T} denotes the tangential projection to the space $m_t(Q)$ at the point $\mathsf{q}(t)$. Recalling that the metric on Q, denoted $\langle\,,\,\rangle_{q(t)}$ is obtained by declaring m_t to be an isometry, the above gives

$$p \cdot w = \langle v + \mathfrak{m}_t^{-1}\mathcal{Z}_t(\mathsf{q}(t))^T, w \rangle_{q(t)} \quad i.e., \quad p = (v + \mathfrak{m}_t^{-1}\mathcal{Z}_t(\mathsf{q}(t))^T)^\flat \tag{6.6.3}$$

where $\flat$ denotes the index lowering operation at $q(t)$ using the metric on Q. The (in general time dependent) Hamiltonian is given by the prescription $H = p \cdot v - L$, which in this case becomes

$$
\begin{aligned}
H_{m_t}(q,p) &= \frac{1}{2}\|p\|^2 - \mathcal{P}(Z_t) - \frac{1}{2}\mathcal{Z}_t^{\perp}\|^2 + V(q) + U(\mathfrak{q}(t)) \\
&= H_0(q,p) - \mathcal{P}(Z_t) - \frac{1}{2}\|\mathcal{Z}_t^{\perp}\|^2 + U(q(t)), \qquad (6.6.4)
\end{aligned}
$$

where $H_0(q,p) = \frac{1}{2}\|p\|^2 + V(q)$, the time dependent vector field $Z_t \in \mathfrak{X}(Q)$ is defined by $Z_t(q) = \mathfrak{m}_t^{-1}[\mathcal{Z}_t(m_t(q))]^T$, the momentum function $\mathcal{P}(Y)$ is defined by $\mathcal{P}(Y)(q,p) = p \cdot Y(q)$ for $Y \in \mathfrak{X}(Q)$, and where $\mathcal{Z}_t^{\perp}$ denotes the orthogonal projection of $\mathcal{Z}_t$ to $m_t(Q)$. Even though the Lagrangian and Hamiltonian are time dependent, we recall that the Euler-Lagrange equations for L_{m_t} are still equivalent to Hamilton's equations for H_{m_t}. These give the correct equations of motion for this moving system. (An interesting example of this is fluid flow on the rotating earth, where it is important to consider the fluid with the motion of the earth superposed, rather than the motion relative to an observer. This point of view is developed in Chern [1991].)

Let G be a Lie group that acts on Q. (For the bead in the hoop, this will be the dynamics of H_0 itself.) We assume for the general theory that H_0 is G-invariant. Assuming the "averaging principle" (*cf.* Arnold [1978], for example) we replace H_{m_t} by its G-average,

$$
\langle H_{m_t}\rangle(q,p) = \frac{1}{2}\|p\|^2 - \langle\mathcal{P}(Z_t)\rangle - \frac{1}{2}\langle\|\mathcal{Z}_t^{\perp}\|^2\rangle + V(q) + \langle U(\mathfrak{q}(t))\rangle \quad (6.6.5)
$$

where $\langle\,.\,\rangle$ denotes the G-average. This principle can be hard to rigorously justify in general. We will use it in a particularly simple example where we will see how to check it directly. Furthermore, we shall discard the term $\frac{1}{2}\langle\|\mathcal{Z}_t^{\perp}\|^2\rangle$; we assume it is small compared to the rest of the terms. Thus, define

$$
\mathcal{H}(q,p,t) = \qquad\qquad\qquad\qquad\qquad\qquad\qquad\qquad\qquad (6.6.6)
$$
$$
\frac{1}{2}\|p\|^2 - \langle\mathcal{P}(Z_t)\rangle + V(q) + \langle U(\mathfrak{q}(t))\rangle = H_0(q,p) - \langle\mathcal{P}(Z_t)\rangle + \langle U(\mathfrak{q}(t))\rangle.
$$

The dynamics of $\mathcal{H}$ on the extended space $T^*Q \times M$ is given by the vector field

$$
(X_{\mathcal{H}}, Z_t) = \left(X_{H_0} - X_{\langle\mathcal{P}(Z_t)\rangle} + X_{\langle U\circ m_t\rangle}, Z_t\right). \qquad (6.6.7)
$$

The vector field

$$
\mathrm{hor}(Z_t) = \left(-X_{\langle\mathcal{P}(Z_t)\rangle}, Z_t\right) \qquad\qquad\qquad (6.6.8)
$$

has a natural interpretation as the horizontal lift of Z_t relative to a connection, which we shall call the **Hannay-Berry connection induced by the Cartan connection**. The holonomy of this connection is interpreted as the Hannay-Berry phase of a slowly moving constrained system.

6.7   The Bead on the Rotating Hoop

Consider Figures 1.5.2 and 1.5.3 which show a hoop (not necessarily circular) on which a bead slides without friction. As the bead is sliding, the hoop is slowly rotated in its plane through an angle $\theta(t)$ and angular velocity $\omega(t) = \dot{\theta}(t)\mathbf{k}$. Let s denote the arc length along the hoop, measured from a reference point on the hoop and let $\mathbf{q}(s)$ be the vector from the origin to the corresponding point on the hoop; thus the shape of the hoop is determined by this function $\mathbf{q}(s)$. Let L be the length of the hoop. The unit tangent vector is $\mathbf{q}'(s)$ and the position of the reference point $\mathbf{q}(s(t))$ relative to an inertial frame in space is $R_{\theta(t)}\mathbf{q}(s(t))$, where R_θ is the rotation in the plane of the hoop through an angle θ.

The configuration space is diffeomorphic to the circle $Q = S^1$. The Lagrangian $L(s, \dot{s}, t)$ is the kinetic energy of the particle; $i.e.$, since

$$\frac{d}{dt}R_{\theta(t)}\mathbf{q}(s(t)) = R_{\theta(t)}\mathbf{q}'(s(t))\dot{s}(t) + R_{\theta(t)}[\omega(t) \times \mathbf{q}(s(t))],$$

we set

$$L(s, \dot{s}, t) = \frac{1}{2}m\|\mathbf{q}'(s)\dot{s} + \omega \times \mathbf{q}(s)\|^2. \tag{6.7.1}$$

The Euler-Lagrange equations

$$\frac{d}{dt}\frac{\partial L}{\partial \dot{s}} = \frac{\partial L}{\partial s}$$

become

$$\frac{d}{dt}m[\dot{s} + \mathbf{q}' \cdot (\omega \times \mathbf{q})] = m[\dot{s}\mathbf{q}'' \cdot (\omega \times \mathbf{q}) + \dot{s}\mathbf{q}' \cdot (\omega \times \mathbf{q}') + (\omega \times \mathbf{q}) \cdot (\omega \times \mathbf{q}')]$$

since $\|\mathbf{q}'\|^2 = 1$. Therefore

$$\ddot{s} + \mathbf{q}'' \cdot (\omega \times \mathbf{q})\dot{s} + \mathbf{q}' \cdot (\dot{\omega} \times \mathbf{q}) = \dot{s}\mathbf{q}'' \cdot (\omega \times \mathbf{q}) + (\omega \times \mathbf{q}) \cdot (\omega \times \mathbf{q}')$$

$i.e.$,

$$\ddot{s} - (\omega \times \mathbf{q}) \cdot (\omega \times \mathbf{q}') + \mathbf{q}' \cdot (\dot{\omega} \times \mathbf{q}) = 0. \tag{6.7.2}$$

The second and third terms in (6.7.2) are the centrifugal and Euler forces respectively. We rewrite (6.7.2) as

$$\ddot{s} = \omega^2 \mathbf{q} \cdot \mathbf{q}' - \dot{\omega}q\sin\alpha \tag{6.7.3}$$

where α is as in Figure 1.5.2 and $q = \|\mathbf{q}\|$. From (6.7.3), Taylor's formula with remainder gives

$$s(t) = s_0 + \dot{s}_0 t + \int_0^t (t - t') \left\{ \omega(t')^2 \mathbf{q} \cdot \mathbf{q}'(s(t')) - \dot{\omega}(t') q(s(t')) \sin \alpha(s(t')) \right\} dt'.$$
(6.7.4)

Now ω and $\dot{\omega}$ are assumed small with respect to the particle's velocity, so by the averaging theorem (see, e.g. Hale [1969]), the s-dependent quantities in (6.7.4) can be replaced by their averages around the hoop:

$$s(T) \approx s_0 + \dot{s}_0 T \qquad\qquad (6.7.5)$$
$$+ \int_0^T (T - t') \left\{ \omega(t')^2 \frac{1}{L} \int_0^L \mathbf{q} \cdot \mathbf{q}' ds - \dot{\omega}(t') \frac{1}{L} \int_0^L q(s) \sin \alpha \, ds \right\} dt'.$$

Aside The essence of the averaging can be seen as follows. Suppose $g(t)$ is a rapidly varying function and $f(t)$ is slowly varying on an interval $[a, b]$. Over one period of g, say $[\alpha, \beta]$, we have

$$\int_\alpha^\beta f(t) g(t) dt \approx \int_\alpha^\beta f(t) \bar{g} dt \qquad\qquad (6.7.6)$$

where

$$\bar{g} = \frac{1}{\beta - \alpha} \int_\alpha^\beta g(t) dt$$

is the average of g. The error in (6.7.6) is

$$\int_\alpha^\beta f(t)(g(t) - \bar{g}) dt$$

which is less than

$$(\beta - \alpha) \times (\text{variation of } f) \times \text{constant} \le \text{constant} \|f'\| (\beta - \alpha)^2.$$

If this is added up over $[a, b]$ one still gets something small as the period of g tends to zero. ♦

The first integral in (6.7.5) over s vanishes and the second is $2A$ where A is the area enclosed by the hoop. Now integrate by parts:

$$\int_0^T (T - t') \dot{\omega}(t') dt' = -T\omega(0) + \int_0^T \omega(t') dt' = -T\omega(0) + 2\pi, \qquad (6.7.7)$$

assuming the hoop makes one complete revolution in time T. Substituting (6.7.7) in (6.7.5) gives

$$s(T) \approx s_0 + \dot{s}_0 T + \frac{2A}{L} \omega_0 T - \frac{4\pi A}{L}. \tag{6.7.8}$$

The initial velocity of the bead *relative to the hoop* is $\dot{s}_0$, while that *relative to the inertial frame* is (see (6.7.1)),

$$v_0 = \mathbf{q}'(0) \cdot [\mathbf{q}'(0)\dot{s}_0 + \omega_0 \times \mathbf{q}(0)] = \dot{s}_0 + \omega_0 q(s_0) \sin \alpha(s_0). \tag{6.7.9}$$

Now average (6.7.8) and (6.7.9) over the initial conditions to get

$$\langle s(T) - s_0 - v_0 T \rangle \approx -\frac{4\pi A}{L} \tag{6.7.10}$$

which means that *on average*, the shift in position is by $4\pi A/L$ between the rotated and nonrotated hoop. This extra length $4\pi A/L$, (or in angular measure, $8\pi^2 A/L^2$) is the Hannay-Berry phase. Note that if $\omega_0 = 0$ (the situation assumed by Berry [1985]) then averaging over initial conditions is not necessary. This process of averaging over the initial conditions used in this example is related to work of Golin and Marmi [1989] on procedures to measure the phase shift.

Chapter 7

Stabilization and Control

In this chapter we present a result of Bloch, Krishnaprasad, Marsden and Sánchez [1991] on the stabilization of rigid body motion using internal rotors followed by a description of some of Montgomery's [1990] work on optimal control.

7.1   The Rigid Body with Internal Rotors

Consider a rigid body (to be called the *carrier body*) carrying one, two or three symmetric rotors. Denote the system center of mass by 0 in the body frame and at 0 place a set of (orthonormal) body axes. Assume that the rotor and the body coordinate axes are aligned with principal axes of the carrier body. The configuration space of the system is $SO(3) \times S^1 \times S^1 \times S^1$.

Let $\mathbb{I}_{\text{body}}$ be the inertia tensor of the carrier, body, $\mathbb{I}_{\text{rotor}}$ the diagonal matrix of rotor inertias about the principal axes and $\mathbb{I}'_{\text{rotor}}$ the remaining rotor inertias about the other axes. Let $\mathbb{I}_{\text{lock}} = \mathbb{I}_{\text{body}} + \mathbb{I}_{\text{rotor}} + \mathbb{I}'_{\text{rotor}}$ be the (body) locked inertia tensor (*i.e.*, with rotors locked) of the full system; this definition coincides with the usage in Chapter 3, except that here the locked inertia tensor is with respect to body coordinates of the carrier body (note that in Chapter 3, the locked inertia tensor is with respect to the *spatial* frame).

The Lagrangian of the free system is the total kinetic energy of the body plus the total kinetic energy of the rotor; *i.e.*,

$$
\begin{aligned}
L &= \frac{1}{2}(\Omega \cdot \mathbb{I}_{\text{body}}\Omega) + \frac{1}{2}\Omega \cdot \mathbb{I}'_{\text{rotor}}\Omega + \frac{1}{2}(\Omega + \Omega_r) \cdot \mathbb{I}_{\text{rotor}}(\Omega + \Omega_r) \\
&= \frac{1}{2}(\Omega \cdot (\mathbb{I}_{\text{lock}} - \mathbb{I}_{\text{rotor}})\Omega) + \frac{1}{2}(\Omega + \Omega_r) \cdot \mathbb{I}_{\text{rotor}}(\Omega + \Omega_r) \quad (7.1.1)
\end{aligned}
$$

where Ω is the vector of body angular velocities and Ω_r is the vector of rotor angular velocities about the principal axes with respect to a (carrier) body fixed frame. Using the Legendre transform, we find the conjugate momenta to be:

$$m = \frac{\partial L}{\partial \Omega} = (\mathbb{I}_{\text{lock}} - \mathbb{I}_{\text{rotor}})\Omega + \mathbb{I}_{\text{rotor}}(\Omega + \Omega_r) = \mathbb{I}_{\text{lock}}\Omega + \mathbb{I}_{\text{rotor}}\Omega_r \quad (7.1.2)$$

and

$$l = \frac{\partial L}{\partial \Omega_r} = \mathbb{I}_{\text{rotor}}(\Omega + \Omega_r) \quad (7.1.3)$$

and the equations of motion including internal torques (controls) u in the rotors are

$$\left. \begin{array}{c} \dot{m} = m \times \Omega = m \times (\mathbb{I}_{\text{lock}} - \mathbb{I}_{\text{rotor}})^{-1}(m - l) \\ \dot{l} = u. \end{array} \right\} \quad (7.1.4)$$

7.2 The Hamiltonian Structure with Feedback Controls

The first result shows that for torques obeying a certain feedback law (*i.e.*, the torques are given functions of the other variables), the preceding set of equations including the internal torques u can still be Hamiltonian! As we shall see, the Hamiltonian structure is of gyroscopic form.

Theorem 7.1 *For the feedback*

$$u = \mathbf{k}(m \times (\mathbb{I}_{\text{lock}} - \mathbb{I}_{\text{rotor}})^{-1}(m - l)), \quad (7.2.1)$$

where $\mathbf{k}$ *is a constant real matrix such that* $\mathbf{k}$ *does not have 1 as an eigenvalue and such that the matrix* $\mathbb{J} = (1 - \mathbf{k})^{-1}(\mathbb{I}_{\text{lock}} - \mathbb{I}_{\text{rotor}})$ *is symmetric, the system (7.1.4) reduces to a Hamiltonian system on* $so(3)^*$ *with respect to the rigid body bracket* $\{F, G\}(m) = -m \cdot (\nabla F \times \nabla G)$.

Proof We have

$$\dot{l} = u = \mathbf{k}\dot{m} = \mathbf{k}((\mathbb{I}_{\text{lock}}\Omega + \mathbb{I}_{\text{rotor}}\Omega_r) \times \Omega). \quad (7.2.2)$$

Therefore, the vector

$$\mathbf{k}m - l = p, \quad (7.2.3)$$

is a constant of motion. Hence our feedback control system becomes

$$\begin{aligned} \dot{m} &= m \times (\mathbb{I}_{\text{lock}} - \mathbb{I}_{\text{rotor}})^{-1}(m - l) \\ &= m \times (\mathbb{I}_{\text{lock}} - \mathbb{I}_{\text{rotor}})^{-1}(m - \mathbf{k}m + p) \\ &= m \times (\mathbb{I}_{\text{lock}} - \mathbb{I}_{\text{rotor}})^{-1}(1 - \mathbf{k})(m - \xi) \end{aligned} \quad (7.2.4)$$

where $\xi = -(1 - \mathbf{k})^{-1}p$ and $\mathbf{1}$ is the identity. Define the $\mathbf{k}$-dependent "inertia tensor"

$$\mathbb{J} = (\mathbf{1} - \mathbf{k})^{-1}(\mathbb{I}_{\text{lock}} - \mathbb{I}_{\text{rotor}}). \tag{7.2.5}$$

Then the equations become

$$\dot{m} = \nabla C \times \nabla H \tag{7.2.6}$$

where

$$C = \frac{1}{2}\|m\|^2 \tag{7.2.7}$$

and

$$H = \frac{1}{2}(m - \xi) \cdot \mathbb{J}^{-1}(m - \xi). \tag{7.2.8}$$

Clearly (7.2.6) are Hamiltonian on $so(3)^*$ with respect to the standard Lie-Poisson structure (see §2.6). ∎

Remark The conservation law (7.2.3), which is key to our methods, may be regarded as a way to choose the control (7.2.1). ♦

Equation (7.2.3) is equivalent to

$$\mathbf{k}(\mathbb{I}_{\text{lock}}\Omega + \mathbb{I}_{\text{rotor}}\Omega_r) - \mathbb{I}_{\text{rotor}}(\Omega + \Omega_r) = p$$

i.e.,

$$(\mathbb{I}_{\text{rotor}} - \mathbf{k}\mathbb{I}_{\text{rotor}})\Omega_r = (\mathbf{k}\mathbb{I}_{\text{lock}} - \mathbb{I}_{\text{rotor}})\Omega - p.$$

Therefore, if

$$\mathbf{k}\mathbb{I}_{\text{lock}} = \mathbb{I}_{\text{rotor}} \tag{7.2.9}$$

then one obtains as a special case, the dual spin case in which the acceleration feedback is such that each rotor rotates at *constant* angular velocity relative to the carrier (see Krishnaprasad [1985] and Sánchez de Alvarez [1986]). Also note that the Hamiltonian in (7.2.8) can be indefinite.

If we set $m_b = m - \xi$, (7.2.4) become

$$\dot{m}_b = (m_b + \xi) \times \mathbb{J}^{-1}m_b. \tag{7.2.10}$$

It is instructive to consider the case where $\mathbf{k} = \text{diag}(k_1, k_2, k_3)$. Let $\mathbb{I} = (\mathbb{I}_{\text{lock}} - \mathbb{I}_{\text{rotor}}) = \text{diag}(\tilde{I}_1, \tilde{I}_2, \tilde{I}_3)$ and the matrix $\mathbb{J}$ satisfies the symmetry hypothesis of Theorem 7.1. Then $l_i = p_i + k_i m_i, i = 1, 2, 3$, and the equations become $\dot{m} = m \times \nabla H$ where

$$H = \frac{1}{2}\left(\frac{((1 - k_1)m_1 + p_1)^2}{(1 - k_1)\tilde{I}_1)} + \frac{((1 - k_2)m_2 + p_2)^2}{(1 - k_2)\tilde{I}_2} + \frac{((1 - k_3)m_3 + p_3)^2}{(1 - k_3)\tilde{I}_3}\right). \tag{7.2.11}$$

It is possible to have more complex feedback mechanisms where the system still reduces to a Hamiltonian system on $so(3)^*$. We refer to Bloch, Krishnaprasad, Marsden and Sánchez [1991] for a discussion of this point.

7.3   Feedback Stabilization of a Rigid Body with a Single Rotor

We now consider the equations for a rigid body with a *single* rotor. We will demonstrate that with a single rotor about the third principal axis, a suitable quadratic feedback stabilizes the system about its intermediate axis.

Let the rigid body have moments of inertia $I_1 > I_2 > I_3$ and suppose the symmetric rotor is aligned with the third principal axis and has moments of inertia $J_1 = J_2$ and J_3. Let $\Omega_i, i = 1, 2, 3$, denote the carrier body angular velocities and let $\dot{\alpha}$ denote that of the rotor (relative to a frame fixed on the carrier body). Let

$$\text{diag}(\lambda_1, \lambda_2, \lambda_3) = \text{diag}(J_1 + I_1, J_2 + I_2, J_3 + I_3) \qquad (7.3.1)$$

be the (body) locked inertia tensor. Then from (7.1.2) and (7.1.3), the natural momenta are

$$\begin{aligned}
m_i &= (J_i + I_i)\Omega_i = \lambda_i \omega_i, i = 1, 2\\
m_3 &= \lambda_3 \Omega_3 + J_3 \dot{\alpha}\\
l_3 &= J_3(\Omega_3 + \dot{\alpha}).
\end{aligned} \qquad (7.3.2)$$

Note that $m_3 = I_3\Omega_3 + l_3$. The equations of motion (7.1.4) are:

$$\begin{aligned}
\dot{m}_1 &= m_2 m_3 \left(\frac{1}{I_3} - \frac{1}{\lambda_2}\right) - \frac{l_3 m_2}{I_3}\\
\dot{m}_2 &= m_1 m_3 \left(\frac{1}{\lambda_1} - \frac{1}{I_3}\right) + \frac{l_3 m_1}{I_3}\\
\dot{m}_3 &= m_1 m_2 \left(\frac{1}{\lambda_2} - \frac{1}{\lambda_1}\right)\\
\dot{l}_3 &= u.
\end{aligned} \qquad (7.3.3)$$

Choosing

$$u = k a_3 m_1 m_2 \quad \text{where} \quad a_3 = \left(\frac{1}{\lambda_2} - \frac{1}{\lambda_1}\right),$$

and noting that $l_3 - k m_3 = p$ is a constant, we get:

Theorem 7.2 *With this choice of u and p, the equations (7.3.3) reduce to*

$$\dot{m}_1 = m_2\left(\frac{(1-k)m_3 - p}{I_3}\right) - \frac{m_3 m_2}{\lambda_2}$$

$$\dot{m}_2 = -m_1\left(\frac{(1-k)m_3 - p}{I_3}\right) + \frac{m_3 m_1}{\lambda_1}$$

$$\dot{m}_3 = a_3 m_1 m_2 \qquad\qquad (7.3.4)$$

which are Hamiltonian on so(3)* with respect to the standard rigid body Lie-Poisson bracket, with Hamiltonian

$$H = \frac{1}{2}\left(\frac{m_1^2}{\lambda_1} + \frac{m_2^2}{\lambda_2} + \frac{((1-k)m_3 - p)^2}{(1-k)I_3)}\right) + \frac{1}{2}\frac{p^2}{J_3(1-k)} \qquad (7.3.5)$$

where p is a constant.

When $k = 0$, we get the equations for the rigid body carrying a free spinning rotor — note that this case is not trivial! The rotor interacts in a nontrivial way with the dynamics of the carrier body. We get the dual spin case for which $J_3\dot{\alpha} = 0$, when $k\mathbb{I}_{\text{lock}} = \mathbb{I}_{\text{rotor}}$ from (7.2.9), or, in this case, when $k = J_3/\lambda_3$. Notice that for this $k, p = (1 - k)\dot{\alpha}$, a multiple of $\dot{\alpha}$.

We can now use the energy-Casimir method to prove

Theorem 7.3 *For $p = 0$ and $k > 1 - (J_3/\lambda_2)$, the system (7.3.4) is stabilized about the middle axis, i.e., about the relative equilibrium $(0, M, 0)$.*

Proof Consider the energy-Casimir function $H + C$ where $C = \varphi(m^2)$, and $m^2 = m_1^2 + m_2^2 + m_3^2$. The first variation is

$$\delta(H + C) = \frac{m_1\delta m_1}{\lambda_1} + \frac{m_2\delta m_2}{\lambda_2} + \frac{(1-k)m_3 - p}{I_3}\delta m_3$$
$$+ \varphi'(m^2)(m_1\delta m_1 + m_2\delta m_2 + m_3\delta m_3). \qquad (7.3.6)$$

This is zero if we choose φ so that

$$\frac{m_1}{\lambda_1} + \varphi' m_1 = 0,$$

$$\frac{m_2}{\lambda_2} + \varphi' m_2 = 0,$$

$$\frac{(1-k)m_3 - p}{I_3} + \varphi' m_3 = 0. \qquad (7.3.7)$$

Then we compute

$$\delta^2(H + C) = \frac{(\delta m_1)^2}{\lambda_1} + \frac{(\delta m_2)^2}{\lambda_2} + \frac{(1-k)(\delta m_3)^2}{I_3}$$
$$+ \varphi'(m^2)((\delta m_1)^2 + (\delta m_2)^2 + (\delta m_3)^2)$$
$$+ \varphi''(m^2)(m_1 \delta m_1 + m_2 \delta m_2 + m_3 \delta m_3)^2. \quad (7.3.8)$$

For $p = 0$, *i.e.*, $l_3 = km_3, (0, M, 0)$ is a relative equilibrium and (7.3.7) are satisfied if $\varphi' = -1/\lambda_2$ at equilibrium. In that case,

$$\delta^2(H + C) = (\delta m_1)^2 \left(\frac{1}{\lambda_1} - \frac{1}{\lambda_2} \right) + (\delta m_3)^2 \left(\frac{1-k}{I_3} - \frac{1}{\lambda_2} \right) + \varphi''(\delta m_2)^2.$$

Now

$$\frac{1}{\lambda_1} - \frac{1}{\lambda_2} = \frac{I_2 - I_1}{\lambda_1 \lambda_2} < 0 \quad \text{for} \quad I_1 > I_2 > I_3.$$

For k satisfying the condition in the theorem,

$$\frac{1-k}{I_3} - \frac{1}{\lambda_2} < 0,$$

so if one chooses φ such that $\varphi'' < 0$ at equilibrium, then the second variation is negative definite and hence stability holds. ∎

For a geometric interpretation of the stabilization found here, see Holm and Marsden [1991], along with other interesting facts about rigid rotors and pendula, for example, a proof that the *rigid body phase space is a union of simple pendulum phase spaces*.

Corresponding to the Hamiltonian (7.3.5) there is a Lagrangian found using the inverse Legendre transformation

$$\tilde{\Omega}_1 = \frac{m_1}{\lambda_1}$$

$$\tilde{\Omega}_2 = \frac{m_2}{\lambda_2}$$

$$\tilde{\Omega}_3 = \frac{(1-k)m_3 - p}{I_3}$$

$$\dot{\tilde{\alpha}} = -\frac{(1-k)m_3 - p}{(1-k)I_3} + \frac{p}{(1-k)I_3}. \quad (7.3.9)$$

Note that $\tilde{\Omega}_1, \tilde{\Omega}_2$ and $\tilde{\Omega}_3$ equal the angular velocities Ω_1, Ω_2, and Ω_3 for the free system, but that $\dot{\tilde{\alpha}}$ is not equal to $\dot{\alpha}$. In fact, we have the interesting *velocity shift*

$$\dot{\tilde{\alpha}} = \frac{\dot{\alpha}}{(1-k)} - \frac{km_3}{(1-k)J_3)}. \quad (7.3.10)$$

Thus the equations on $TSO(3)$ determined by (7.3.9) are the Euler-Lagrange equations for a Lagrangian quadratic in the velocities, so the equations can be regarded as geodesic equations. The torques can be thought of as residing in the velocity shift (7.3.10). Using the free Lagrangian, the torques appear as generalized forces on the right hand side of the Euler-Lagrange equations. Thus, the d'Alembert principle can be used to describe the Euler-Lagrange equations with the generalized forces. This approach arranges in a different way, the useful fact that the equations are derivable from a Lagrangian (and hence a Hamiltonian) in velocity shifted variables. In fact, it seems that the right context for this is the Routhian and the theory of gyroscopic Lagrangians, as in §3.6, but we will not pursue this further here.

For problems like the driven rotor and specifically the dual spin case where the rotors are driven with constant angular velocity, one might think that this is a velocity constraint and should be treated by using constraint theory. For the particular problem at hand, this can be circumvented and in fact, standard methods are applicable and constraint theory is not needed, as we have shown.

7.4 Phase Shifts

Next, we discuss an attitude drift that occurs in the system and suggest a method for correcting it. If the system (7.2.10) is perturbed from a stable equilibrium, and the perturbation is not too large, the closed loop system executes a periodic motion on a level surface (momentum sphere) of the Casimir function $\|m_b + \xi\|^2$ in the body-rotor feedback system. This leads to an attitude drift which can be thought of as rotation about the (constant) spatial angular momentum vector. We calculate the amount of this rotation following the method of Montgomery we used in Chapter 6 to calculate the phase shift for the single rigid body.

As we have seen, the equations of motion for the rigid body-rotor system with feedback law (7.2.1) are

$$\dot{m}_b = (m_b + \xi) \times \mathbb{J}^{-1} m_b = \nabla C \times \nabla H \qquad (7.4.1)$$

where $m_b = m + (1 - \mathbf{k})^{-1}(\mathbf{k}m - l), \xi = -(1 - \mathbf{k})^{-1}(\mathbf{k}m - l)$, and

$$C = \frac{1}{2}\|m_b + \xi\|^2 \qquad (7.4.2)$$

$$H = \frac{1}{2}m_b \cdot \mathbb{J}^{-1} m_b. \qquad (7.4.3)$$

As with the rigid body, this system is completely integrable with trajectories given by intersecting the level sets of C and H. Note that (7.4.2) just defines a sphere shifted by the amount ξ.

The attitude equation for the rigid-body rotor system is

$$\dot{A} = A\hat{\Omega},$$

where $\hat{}$ denotes the isomorphism between $\mathbb{R}^3$ and $so(3)$ (see §4.4) and in the presence of the feedback law (7.2.1),

$$
\begin{aligned}
\Omega &= (\mathbb{I}_{\text{lock}} - \mathbb{I}_{\text{rotor}})^{-1}(m - l) \\
&= (\mathbb{I}_{\text{lock}} - \mathbb{I}_{\text{rotor}})^{-1}(m - \mathbf{k}m + p) \\
&= \mathbb{J}^{-1}(m - \xi) = \mathbb{J}^{-1}m_b.
\end{aligned}
$$

Therefore the attitude equation may be written

$$\dot{A} = A(\mathbb{J}^{-1}m_b)\hat{}. \tag{7.4.4}$$

The net spatial (constant) angular momentum vector is

$$\mu = A(m_b + \xi). \tag{7.4.5}$$

Then we have the following:

Theorem 7.4 *Suppose the solution of*

$$\dot{m}_b = (m_b + \xi) \times \mathbb{J}^{-1}m_b \tag{7.4.6}$$

is a periodic orbit of period T on the momentum sphere, $\|m_b + \xi\|^2 = \|\mu\|^2$, enclosing a solid angle Φ_{solid}. Let Ω_{av} denote the average value of the body angular velocity over this period, E denote the constant value of the Hamiltonian, and $\|\mu\|$ denote the magnitude of the angular momentum vector. Then the body undergoes a net rotation $\Delta\theta$ about the spatial angular momentum vector μ given by

$$\Delta\theta = \frac{2ET}{\|\mu\|} + \frac{T}{\|\mu\|}(\xi \cdot \Omega_{\text{av}}) - \Phi_{\text{solid}}. \tag{7.4.7}$$

Proof Consider the reduced phase space (the momentum sphere)

$$P_\mu = \{m_b \mid \|m_b + \xi\|^2 = \|\mu\|^2\}, \tag{7.4.8}$$

for μ fixed, so that in this space,

$$m_b(t_0 + T) = m_b(t_0), \tag{7.4.9}$$

and from momentum conservation

$$\mu = A(T + t_0)(m_b(T + t_0) + \xi) = A(t_0)(m_b(t_0) + \xi). \qquad (7.4.10)$$

Hence

$$A(T + t_0)A(t_0)^{-1}\mu = \mu.$$

Thus $A(T + t_0)A(t_0)^{-1} \in G_\mu$, so

$$A(T + t_0)A(t_0)^{-1} = \exp\left(\Delta\theta \frac{\mu}{\|\mu\|}\right) \qquad (7.4.11)$$

for some $\Delta\theta$, which we wish to determine.

We can assume that $A(t_0)$ is the identity, so the body is in the reference configuration at $t_0 = 0$ and thus that $m_b(t_0) = \mu - \xi$. Consider the phase space trajectory of our system

$$z(t) = (A(t), m_b(t)), \quad z(t_0) = z_0. \qquad (7.4.12)$$

The two curves in phase space

$$C_1 = \{z(t) \mid t_0 \le t \le t_0 + T\}$$

(the dynamical evolution from z_0), and

$$C_2 = \left\{ \exp\left(\theta \frac{\mu}{\|\mu\|}\right) z_0 \;\Bigg|\; 0 \le \theta \le \Delta\theta \right\}$$

intersect at $t = T$. Thus $C = C_1 - C_2$ is a closed curve in phase space and so by Stokes' theorem,

$$\int_{C_1} p\, dq - \int_{C_2} p\, dq = \iint_\Sigma d(p\, dq) \qquad (7.4.13)$$

where $p\, dq = \sum_{i=1}^{3} p_i dq_i$ and where q_i and p_i are configuration space variables and conjugate momenta in the phase space and Σ is a surface enclosed by the curve C. Evaluating each of these integrals will give the formula for $\Delta\theta$. Letting ω be the spatial angular velocity, we get

$$p\frac{dq}{dt} = \mu \cdot \omega = A(\mathbb{J}\Omega + \xi) \cdot A\Omega = \mathbb{J}\Omega \cdot \Omega + \xi \cdot \Omega. \qquad (7.4.14)$$

Hence

$$\int_{C_1} p\, dq = \int_{C_1} p \cdot \frac{dq}{dt} dt = \int_0^T \mathbb{J}\Omega \cdot \Omega dt + \int_0^T \xi \cdot \Omega dt = 2ET + (\xi \cdot \Omega_{\text{av}})T \qquad (7.4.15)$$

since the Hamiltonian is conserved along orbits. Along C_2,

$$\int_{C_2} p \cdot \frac{dq}{dt} dt = \int_{C_2} \mu \cdot \omega dt = \int_{C_2} \mu \cdot \left\{ \frac{d\theta}{dt} \frac{\mu}{\|\mu\|} \right\} dt = \|\mu\| \int_{C_2} d\theta = \|\mu\| \Delta\theta.$$
(7.4.16)

Finally we note that the map π_μ from the set of points in phase space with angular momentum μ to P_μ satisfies

$$\iint_\Sigma d(pdq) = \iint_{\pi_\mu(\Sigma)} dA = \|\mu\| \Phi_{\text{solid}}$$
(7.4.17)

where dA is the area form on the two-sphere and $\pi_\mu(\Sigma)$ is the spherical cap bounded by the periodic orbit $\{m_b(t) \mid t_0 \leq t \leq t_0 + T\} \subset P_\mu$. Combining (7.4.15) – (7.4.17) we get the result. ■

Remarks

1 When $\xi = 0$, (7.4.7) reduces to Montgomery's formula in Chapter 6.

2 This theorem may be viewed as a special case of a scenario that is useful for other systems, such as rigid bodies with flexible appendages. As we saw in Chapter 6, phases may be viewed as occurring in the reconstruction process, which lifts the dynamics from P_μ to $\mathbf{J}^{-1}(\mu)$. By the cotangent bundle reduction theorem, P_μ is a bundle over T^*S, where $S = Q/G$ is shape space. The fiber of this bundle is O_μ, the coadjoint orbit through μ. For a rigid body with three internal rotors, S is the three torus $\mathbb{T}^3$ parametrized by the rotor angles. Controlling them by a feedback or other control and using other conserved quantities associated with the rotors as we have done, leaves one with dynamics on the "rigid variables" O_μ, the momentum sphere in our case. Then the problem reduces to that of lifting the dynamics on O_μ to $\mathbf{J}^{-1}(\mu)$ with the T^*S dynamics given. For $G = SO(3)$ this "reduces" the problem to that for geometric phases for the rigid body.

3 Some interesting control manouvers for "satellite parking" using these ideas may be found in Walsh [1991].

4 See Montgomery [1990], p. 569 for some insightful comments on the Chow-Ambrose-Singer theorem in this context. ◆

Finally, following a suggestion of Krishnaprasad, we show that in the zero total angular momentum case one can compensate for this drift using *two* rotors. The total *spatial* angular momentum if one has only two rotors is of the form

$$\mu = A(\mathbb{I}_{\text{lock}}\Omega + b_1 \dot{\alpha}_1 + b_2 \dot{\alpha}_2)$$
(7.4.18)

where the scalars $\dot{\alpha}_1$ and $\dot{\alpha}_2$ represent the rotor velocities relative to the body frame. The attitude matrix A satisfies

$$\dot{A} = A\hat{\Omega}, \tag{7.4.19}$$

as above. If $\mu = 0$, then from (7.4.18) and (7.4.19) we get,

$$\dot{A} = -A((\mathbb{I}_{\text{lock}}^{-1}b_1)\dot{\alpha}_1 + (\mathbb{I}_{\text{lock}}^{-1}b_2)\dot{\alpha}_2). \tag{7.4.20}$$

It is well known (see for instance Brockett [1973] or Crouch [1986]) that if we treat the $\dot{\alpha}_i, i = 1, 2$ as *controls*, then attitude controllability holds iff

$$(\mathbb{I}_{\text{lock}}^{-1}b_1)^{\hat{}} \quad \text{and} \quad (\mathbb{I}_{\text{lock}}^{-1}b_2)^{\hat{}} \quad \text{generate} \quad so(3),$$

or equivalently, iff the vectors

$$\mathbb{I}_{\text{lock}}^{-1}b_1 \quad \text{and} \quad \mathbb{I}_{\text{lock}}^{-1}b_2 \quad \text{are linearly independent.} \tag{7.4.21}$$

Moreover, one can write the attitude matrix as a reverse path-ordered exponential

$$A(t) = A(0) \cdot \bar{\mathbb{P}} \exp\left[-\int_0^t \left\{(\mathbb{I}_{\text{lock}}^{-1}b_1)\dot{\alpha}_1(\sigma) + (\mathbb{I}_{\text{lock}}^{-1}b_2)\dot{\alpha}_2(\sigma)\right\} d\sigma\right]. \tag{7.4.22}$$

The right hand side of (7.4.22) depends only on the *path traversed* in the space $\mathbb{T}^2$ of rotor angles (α_1, α_2) and *not* on the history of velocities $\dot{\alpha}_i$. Hence the formula (7.4.22) should be interpreted as a "geometric phase". Furthermore, the controllability condition can be interpreted as a curvature condition on the principal connection on the bundle $\mathbb{T}^2 \times SO(3) \to \mathbb{T}^2$ defined by the $so(3)$-valued differential 1-form,

$$\theta(\alpha_1, \alpha_2) = -((\mathbb{I}_{\text{lock}}^{-1}b_1)d\alpha_1 + (\mathbb{I}_{\text{lock}}^{-1}b_2)d\alpha_2). \tag{7.4.23}$$

For further details on this geometric picture of multibody interaction see Krishnaprasad [1990], Krishnaprasad and Wang [1992], and for use of these ideas in understanding how to "park a satellite", see Walsh [1991]. In fact, our discussion of the two coupled bodies in §6.1 can be viewed as an especially simple planar version of what is required.

7.5 The Kaluza-Klein Description of Charged Particles

In preparation for the next section we describe the equations of a charged particle in a magnetic field in terms of geodesics. The description we saw in §2.10 can be obtained from the Kaluza-Klein description using an S^1-reduction. The process described here generalizes to the case of a particle in a Yang-Mills field by replacing the magnetic potential A by the Yang-Mills connection.

We are motivated as follows: since charge is a conserved quantity, we introduce a new cyclic variable whose conjugate momentum is the charge. This process is applicable to other situations as well; for example, in fluid dynamics one can profitably introduce a variable conjugate to the conserved mass density or entropy; *cf.* Marsden, Ratiu and Weinstein [1984a,b]. For a charged particle, the resultant system is in fact geodesic motion!

Recall that if $\mathbf{B} = -\nabla \times \mathbf{A}$ is a given magnetic field on $\mathbb{R}^3$, then with respect to canonical variables $(\mathbf{q}, \mathbf{p})$, the Hamiltonian is

$$H(\mathbf{q}, \mathbf{p}) = \frac{1}{2m} \|\mathbf{p} - \frac{e}{c}\mathbf{A}\|^2. \tag{7.5.1}$$

We can obtain (7.5.1) via the Legendre transform if we choose

$$L(\mathbf{q}, \dot{\mathbf{q}}) = \frac{1}{2}m\|\dot{\mathbf{q}}\|^2 + \frac{e}{c}\mathbf{A} \cdot \dot{\mathbf{q}} \tag{7.5.2}$$

for then

$$\mathbf{p} = \frac{\partial L}{\partial \dot{\mathbf{q}}} = m\dot{\mathbf{q}} + \frac{e}{c}\mathbf{A} \tag{7.5.3}$$

and

$$\begin{aligned}
\mathbf{p} \cdot \dot{\mathbf{q}} - L(\mathbf{q}, \dot{\mathbf{q}}) &= (m\dot{\mathbf{q}} + \frac{e}{c}\mathbf{A}) \cdot \dot{\mathbf{q}} - \frac{1}{2}m\|\dot{\mathbf{q}}\|^2 - \frac{e}{c}\mathbf{A} \cdot \dot{\mathbf{q}} \\
&= \frac{1}{2}m\|\dot{\mathbf{q}}\|^2 \\
&= \frac{1}{2m}\|\mathbf{p} - \frac{e}{c}\mathbf{A}\|^2 = H(\mathbf{q}, \mathbf{p}).
\end{aligned} \tag{7.5.4}$$

Thus, the Euler-Lagrange equations for (7.5.2) reproduce the equations for a particle in a magnetic field. (If an electric field $\mathbf{E} = -\nabla\varphi$ is present as well, subtract $e\varphi$ from L, treating $e\varphi$ as a potential energy.) Let the **Kaluza-Klein configuration space** be

$$Q_K = \mathbb{R}^3 \times S^1 \tag{7.5.5}$$

with variables $(\mathbf{q}, \theta)$ and consider the one-form

$$\omega = A + \mathbf{d}\theta \tag{7.5.6}$$

on Q_K regarded as a connection one-form. Define the **Kaluza-Klein Lagrangian** by

$$
\begin{aligned}
L_K(\mathbf{q}, \dot{\mathbf{q}}, \theta, \dot{\theta}) &= \frac{1}{2}m\|\dot{\mathbf{q}}\|^2 + \frac{1}{2}\|\langle\omega, (\mathbf{q}, \dot{\mathbf{q}}, \theta, \dot{\theta})\rangle\|^2 \\
&= \frac{1}{2}m\|\dot{\mathbf{q}}\|^2 + \frac{1}{2}(\mathbf{A}\cdot\dot{\mathbf{q}} + \dot{\theta})^2.
\end{aligned}
\tag{7.5.7}
$$

The corresponding momenta are

$$
\mathfrak{p} = m\dot{\mathbf{q}} + (\mathbf{A}\cdot\dot{\mathbf{q}} + \dot{\theta})\mathbf{A} \quad \text{and} \quad p_\theta = \mathbf{A}\cdot\dot{\mathbf{q}} + \dot{\theta}.
\tag{7.5.8}
$$

Since (7.5.7) is quadratic and positive definite in $\dot{\mathbf{q}}$ and $\dot{\theta}$, *the Euler-Lagrange equations are the geodesic equations on* $\mathbb{R}^3 \times S^1$ *for the metric for which* L_K *is the kinetic energy.* Since p_θ is constant in time as can be seen from the Euler-Lagrange equation for $(\theta, \dot{\theta})$, we can define the **charge** e by setting

$$
p_\theta = e/c;
\tag{7.5.9}
$$

then (7.5.8) coincides with (7.5.3). The corresponding Hamiltonian on T^*Q_K endowed with the canonical symplectic form is

$$
H_K(\mathbf{q}, \mathfrak{p}, \theta, p_\theta) = \frac{1}{2}m\|\mathfrak{p} - p_\theta A\|^2 + \frac{1}{2}p_\theta^2.
\tag{7.5.10}
$$

Since p_θ is constant, H_K differs from H only by the constant $p_\theta^2/2$.

These constructions generalize to the case of a particle in a Yang-Mills field where ω becomes the connection of a Yang-Mills field and its curvature measures the field strength which, for an electromagnetic field, reproduces the relation $\mathbf{B} = \nabla \times \mathbf{A}$. We refer to Wong [1970], Sternberg [1977], Weinstein [1978] and Montgomery [1985] for details and further references. Finally, we remark that the relativistic context is the most natural to introduce the full electromagnetic field. In that setting the construction we have given for the magnetic field will include both electric and magnetic effects. Consult Misner, Thorne and Wheeler [1972] and Gotay et al. [1992] for additional information.

Notice that the reduction of the Kaluza-Klein system by S^1 reproduces the description in terms of (7.5.1) or (7.5.2). The magnetic terms in the sense of reduction become the magnetic terms we started with. Note that the description in terms of the Routhian from §3.6 can also be used here, reproducing the same results. Also notice the similar way that the gyroscopic terms enter into the rigid body system with rotors — they too can be viewed as magnetic terms obtained through reduction.

7.6   Optimal Control and Yang-Mills Particles

In this section we briefly discuss an elegant link between optimal control and the dynamics of a particle in a Yang-Mills field that was discovered by Wilczek, Shapere and Montgomery. We refer to Montgomery [1990] for further details and references. This topic, together with the use of connections described in previous chapters has lead one to speak about the "gauge theory of deformable bodies". The example of a falling cat as a control system is good to keep in mind while reading this section.

We start with a configuration space Q and assume we have a symmetry group G acting freely by isometries, as before. Put on the bundle $Q \to S = Q/G$ the mechanical connection, as described in §3.3. Fix a point $q_0 \in Q$ and a group element $g \in G$. Let

$$\text{hor}(q_0, gq_0) = \text{ all horizontal paths joining } q_0 \text{ to } gq_0. \tag{7.6.1}$$

This space of (suitably differentiable) paths may be regarded as the space of horizontal paths with a *given holonomy* g. Recall from §3.3 that horizontal means that this path is one along which the total angular momentum (*i.e.*, the momentum map of the curve in T^*Q corresponding to $\dot{q}$) is zero.

The projection of the curves in hor (q_0, gq_0) to shape space $S = Q/G$ are closed, as in Figure 7.6.1.

The optimal control problem we wish to discuss is: *Find a path c in* hor(q_0, gq_0) *whose base curve s has minimal (or extremal) length.* The minimum (or extremum) is taken over all paths s obtained from projections of paths in hor(q_0, gq_0).

One may wish to use functions other than the length of c to extremize. For example, the cat may wish to turn itself over by minimizing the amount of work done.

The ***Wong Hamiltonian*** $H_W : T^*Q \to \mathbb{R}$ is defined by

$$H_W(q, p) = \frac{1}{2}\|\text{hor}(q, p)\|^2 \tag{7.6.2}$$

where hor is the horizontal projection defined in §3.3. The function H_W may also be regarded as a function on T^*S — it is the kinetic energy associated to the metric induced on S and as such, its integral curves are solutions of ***Wong's equations***; note that (7.6.2) corresponds to (7.5.1) and A is replaced by the mechanical connection. The corresponding Kaluza-Klein metric on Q is the given metric and its Hamiltonian is the Wong Hamiltonian plus $\frac{1}{2}\|\mathbf{J}\|^2$, assuming that $\mathfrak{g}$ carries an Ad-invariant metric. The situation of the preceding section then becomes a special case of this one.

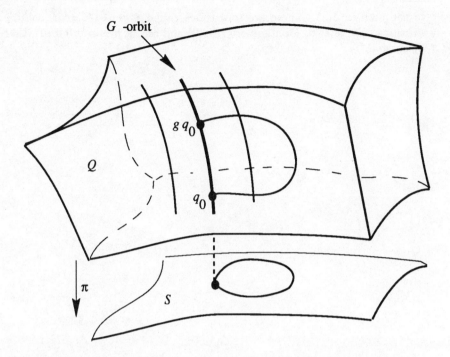

Figure 7.6.1: Loops in Q with a fixed holonomy g and base point q_0.

Theorem 7.5 *Consider a path* $c \in \mathrm{hor}(q_0, gq_0)$ *with given holonomy* g. *Then* c *is extremal iff the curve* s *satisfies Wong's equations.*

This result is closely related to, and may be deduced from, our work on Lagrangian reduction in §3.6. The point is that $c \in \mathrm{hor}(q_0, gq_0)$ means $\dot{c} \in \mathbf{J}^{-1}(0)$, and what we want to do is relate a variational problem on Q but within $\mathbf{J}^{-1}(0)$ and with Hamiltonian the Kaluza-Klein kinetic energy on TQ with a reduced variational principle on S. This is exactly the set up to which the discussion in §3.6 applies.

We also note that s is the projection of a *geodesic* in Q in $\mathrm{hor}(q_0, gq_0)$ obtained from c by adjusting the phase, as in Chapter 6.

What is less obvious is how to use the methods of the calculus of variations to show the existence of minimizing loops with a given holonomy and what their smoothness properties are. For this purpose, the subject of sub-Riemannian geometry (dealing with degenerate metrics) is relevant. Note that H_W in (7.6.2) when regarded as a Hamiltonian on T^*Q (rather that T^*S) is associated with a degenerate co-metric (a degenerate bilinear form in the momentum). These are sometimes called ***Carnot-Caratheodory metrics***. We refer to Montgomery [1990] (and recent preprints) for further discussion.

Chapter 8

Discrete reduction

In this chapter, we extend the theory of reduction of Hamiltonian systems with symmetry to include systems with a discrete symmetry group acting *symplectically*. The exposition here is based on the work of Harnad, Hurtubise and Marsden [1991].

For *antisymplectic* symmetries such as reversibility, this question has been considered by Meyer [1981] and Wan [1990]. However, in this chapter we are concerned with *symplectic* symmetries. Antisymplectic symmetries are typified by time reversal symmetry, while symplectic symmetries are typified by spatial discrete symmetries of systems like reflection symmetry. Often these are obtained by taking the cotangent lift of a discrete symmetry of configuration space.

There are two main motivations for the study of discrete symmetries. The first is the theory of bifurcation of relative equilibria in mechanical systems with symmetry. The rotating liquid drop is a system with a symmetric relative equilibrium that bifurcates via a discrete symmetry. An initially circular drop (with symmetry group S^1) that is rotating rigidly in the plane with constant angular velocity Ω, radius r, and with surface tension τ, is stable if $r^3\Omega^2 < 12\tau$ (this is proved by the energy-Casimir or energy-momentum method). Another relative equilibrium (a rigidly rotating solution in this example) branches from this circular solution at the critical point $r^3\Omega^2 = 12\tau$. The new solution has the spatial symmetry of an ellipse; that is, it has the symmetry $\mathbb{Z}_2 \times \mathbb{Z}_2$ (or equivalently, the dihedral group $\mathbb{D}_2$). These new solutions are stable, although whether they are subcritical or supercritical depends on the parametrization chosen (angular velocity vs. angular momentum, for example). This example is taken from Lewis, Marsden and Ratiu [1987] and Lewis [1989]. It also motivated some of the work in the general theory of bifurcation of equilibria for Hamiltonian systems with symmetry by Golubitsky and Stewart [1987].

The *spherical pendulum* is an especially simple mechanical system having both continuous (S^1) and discrete $(\mathbb{Z}_2)$ symmetries. Solutions can be relative equilibria for the action of rotations about the axis of gravity in which the pendulum rotates in a circular motion. For zero angular momentum these solutions degenerate to give planar oscillations, which lie in the fixed point set for the action of reflection in that plane. In our approach, this reflection symmetry is cotangent lifted to produce a symplectic involution of phase space. The invariant subsystem defined by the fixed point set of this involution is just the *planar pendulum*.

The *double spherical pendulum* has the same continuous and discrete symmetry group. However, the double spherical pendulum has another nontrivial symmetry involving a spacetime symmetry, much as the rotating liquid drop and the water molecule. Namely, we are at a fixed point of a symmetry when the two pendula are rotating in a steadily rotating vertical plane — the symmetry is reflection in this plane. This symmetry is, in fact, the source of the subblocking property of the second variation of the amended potential that we observed in Chapter 5. In the S^1 reduced space, a steadily rotating plane is a stationary plane and the discrete symmetry just becomes reflection in that plane.

A related example is the *double whirling mass system*. This consists of two masses connected to each other and to two fixed supports by springs, but with no gravity. Assume the masses and springs are identical. Then there are two $\mathbb{Z}_2$ symmetries now, corresponding to reflection in a (steadily rotating) vertical plane as with the double spherical pendulum, and to swapping the two masses in a horizontal plane.

The *classical water molecule* has a discrete symplectic symmetry group $\mathbb{Z}_2$ as well as the continuous symmetry group $SO(3)$. The discrete symmetry is closely related to the symmetry of exchanging the two hydrogen atoms.

For these examples, there are some basic links to be made with the block diagonalization work from Chapter 5. In particular, we show how the discrete symmetry can be used to refine the block structure of the second variation of the augmented Hamiltonian and of the symplectic form. Recall that the block diagonalization method provides coordinates in which the second variation of the amended potential on the reduced configuration space is a block diagonal matrix, with the group variables separated from the internal variables; the group part corresponds to a bilinear form computed first by Arnold for purposes of examples whose configuration space is a group. The internal part corresponds to the shape space variables, that are on Q/G the quotient of configuration space by the continuous symmetry group. For the water molecule, we find that the discrete symmetry provides a further blocking of this second variation, by splitting the internal tangent space naturally into symmetric modes and nonsymmetric ones.

In the *dynamics of coupled rigid bodies* one has interesting symmetry breaking bifurcations of relative equilibria and of relative periodic orbits. For the latter, discrete spacetime symmetries are important. We refer to Montaldi, Stewart and Roberts [1988], Oh, Sreenath, Krishnaprasad and Marsden [1989] and to Patrick [1989, 1990] for further details. In the optimal control problem of the falling cat (Montgomery [1990] and references therein), the problem is modeled as the dynamics of two identical coupled rigid bodies and the fixed point set of the involution that swaps the bodies describes the "no-twist" condition of Kane and Shur [1969]; this plays an essential role in the problem and the dynamics is integrable on this set.

The second class of examples motivating the study of discrete symmetries are integrable systems, including those of Bobenko, Reyman and Semenov-Tian-Shansky [1989]. This (spectacular) reference shows in particular how reduction and dual pairs, together with the theory of R-matrices, can be used to understand the integrability of a rich class of systems, including the celebrated Kowalewski top. They also obtain all of the attendant algebraic geometry in this context. It is clear from their work that discrete symmetries, and specifically those obtained from Cartan involutions, play a crucial role. We refer the reader to the paper of Harnad, Hurtubise and Marsden [1991] for the details of this topic.

8.1 Fixed Point Sets and Discrete Reduction

Let (P, Ω) be a symplectic manifold, G a Lie group acting on P by symplectic transformations and $\mathbf{J} : P \to \mathfrak{g}^*$ an Ad^*-equivariant momentum map for the G-action. Let Σ be a compact Lie group acting on P by symplectic transformations and by group homomorphisms on G. For $\sigma \in \Sigma$, write

$$\sigma_P : P \to P \quad \text{and} \quad \sigma_G : G \to G$$

for the corresponding symplectic map on P and group homomorphism on G. Let $\sigma_{\mathfrak{g}} : \mathfrak{g} \to \mathfrak{g}$ be the induced Lie algebra homomorphism (the derivative of σ_G at the identity) and $\sigma_{\mathfrak{g}^*} : \mathfrak{g}^* \to \mathfrak{g}^*$ be the dual of $(\sigma^{-1})_{\mathfrak{g}}$, so $\sigma_{\mathfrak{g}^*}$ is a Poisson map with respect to the Lie-Poisson structure.

Remarks

1 One can also consider the Poisson case directly using the methods of Poisson reduction (Marsden and Ratiu [1986]) or by considering the present work applied to their symplectic leaves.

2 As in the general theory of equivariant momentum maps, one may drop the equivariance assumption by using another action on $\mathfrak{g}^*$.

3 Compactness of Σ is used to give invariant metrics obtained by averaging over Σ — it can be weakened to the existence of invariant finite measures on Σ.

4 The G-action will be assumed to be a left action, although right actions can be treated in the same way. ♦

Assumption 1 *The actions of G and of Σ are compatible in the sense that the following equation holds:*

$$\sigma_P \circ g_P = [\sigma_G(g)]_P \circ \sigma_P \qquad (8.1.1)$$

for each $\sigma \in \Sigma$ and $g \in G$, where we have written g_P for the action of $g \in G$ on P. (See Figure 8.1.1.)

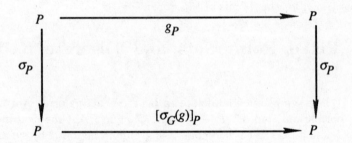

Figure 8.1.1: Assumption 1.

If we differentiate equation (8.1.1) with respect to g at the identity $g = e$, in the direction $\xi \in \mathfrak{g}$, we get

$$T\sigma_P \circ X_{\langle J,\xi \rangle} = X_{\langle J,\sigma_{\mathfrak{g}} \cdot \xi \rangle} \circ \sigma_P \qquad (8.1.2)$$

where X_f is the Hamiltonian vector field on P generated by the function $f : P \to \mathbb{R}$. Since σ_P is symplectic, (8.1.2) is equivalent to

$$X_{\langle J,\xi \rangle \circ \sigma_P^{-1}} = X_{\langle J,\sigma_{\mathfrak{g}} \cdot \xi \rangle} \qquad (8.1.3)$$

i.e.,

$$\mathbf{J} \circ \sigma_P = \sigma_{\mathfrak{g}^*} \circ \mathbf{J} + (\text{cocycle}). \tag{8.1.4}$$

We shall assume, in addition, that the cocycle is zero. In other words, we make:

Assumption 2 *The following equation holds (see Figure 8.1.2):*

$$\mathbf{J} \circ \sigma_P = \sigma_{\mathfrak{g}^*} \circ \mathbf{J}. \tag{8.1.5}$$

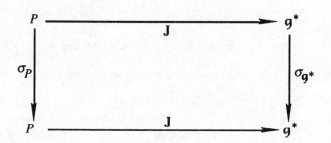

Figure 8.1.2: Assumption 2.

Let $G_\Sigma = \text{Fix}(\Sigma, G) \subset G$ be the fixed point set of Σ; that is,

$$G_\Sigma = \{g \in G \,|\, \sigma_G(g) = g \quad \text{for all} \quad \sigma \in \Sigma\}. \tag{8.1.6}$$

The Lie algebra of G_Σ is the fixed point (*i.e.*, eigenvalue 1) subspace:

$$\mathfrak{g}_\Sigma = \text{Fix}(\Sigma, \mathfrak{g}) = \{\xi \in \mathfrak{g} \,|\, T\sigma_G(e) \cdot \xi = \xi \quad \text{for all} \quad \sigma \in \Sigma\}. \tag{8.1.7}$$

Remarks

1 If G is connected, then (8.1.5) (or (8.1.4)) implies (8.1.1).

2 If Σ is a discrete group, then Assumptions 1 and 2 say that $\mathbf{J} : P \to \mathfrak{g}^*$ is an equivariant momentum map for the semi-direct product $\Sigma \circledS G$, which has the multiplication

$$(\sigma_1, g_1) \cdot (\sigma_2, g_2) = (\sigma_1 \sigma_2, (\sigma_2)_G(g_1)g_2). \tag{8.1.8}$$

Recall that Σ embeds as a subgroup of $\Sigma \circledS G$ via $\sigma \mapsto (\sigma, e)$. The action of Σ on G is given by conjugation within $\Sigma \circledS G$. If one prefers, one can take the point of view that one starts with a group $\mathcal{G}$ with G chosen to be the connected component of the identity and with Σ a subgroup of $\mathcal{G}$ isomorphic to the quotient of $\mathcal{G}$ by the normal subgroup G.

3 One checks that $\Sigma \circledS G_\Sigma = N(\Sigma)$, where $N(\Sigma)$ is the normalizer of $\Sigma \times \{e\}$ within $\mathcal{G} = \Sigma \circledS G$. Accordingly, we can identify G_Σ with the quotient $N(\Sigma)/\Sigma$. In Golubitsky and Stewart [1987] and related works, the tendency is to work with the group $N(\Sigma)/\Sigma$; it will be a bit more convenient for us to work with the group G_Σ, although the two approaches are equivalent, as we have noted.   ♦

Let $P_\Sigma = \mathrm{Fix}(\Sigma, P) \subset P$ be the fixed point set for the action of Σ on P:

$$P_\Sigma = \{z \in P \,|\, \sigma_P(z) = z \quad \text{for all} \quad \sigma \in \Sigma\}. \tag{8.1.9}$$

Proposition 8.1 *The fixed point set P_Σ is a smooth symplectic submanifold of P.*

Proof Put on P a metric that is Σ-invariant and exponentiate the linear fixed point set $(T_z P)_\Sigma$ of the tangent map $T_z \sigma : T_z \sigma : T_z P \to T_z P$ for each $z \in P_\Sigma$ to give a local chart for P_Σ. Since $(T_z P)_\Sigma$ is invariant under the associated complex structure, it is symplectic, so P_Σ is symplectic.   ∎

Proposition 8.2 *The manifold P_Σ is invariant under the action of G_Σ.*

Proof Let $z \in P_\Sigma$ and $g \in G_\Sigma$. To show that $g_P(z) \in P_\Sigma$, we show that $\sigma_P(g_P(z)) = g_P(z)$ for $\sigma \in \Sigma$. To see this, note that by (8.1.1), and the facts that $\sigma_G(g) = g$ and $\sigma_P(z) = z$,

$$\sigma_P(g_P(z)) = [\sigma_G(g)]_P(\sigma_P(z)) = g_P(z). \quad ∎$$

The main message of this section is the following:

Discrete Reduction Procedure *If H is a Hamiltonian system on P that is G and Σ invariant, then its Hamiltonian vector field X_H (or its flow) will leave P_Σ invariant and so standard symplectic reduction can be performed with respect to the action of the symmetry group G_Σ on P_Σ.*

We are not necessarily advocating that one should *always* perform this discrete reduction procedure, or that it contains in some sense equivalent information to the original system, as is the case with continuous reduction.

We *do* claim that the procedure is an interesting way to identify invariant subsystems, and to generate new ones that are important in their own right. For example, reducing the spherical pendulum by discrete reflection in a plane gives the half-dimensional simple planar pendulum. Discrete reduction of the double spherical pendulum by discrete symmetry produces a planar compound pendulum, possibly with magnetic terms. More impressively, the Kowalewski top arises by discrete reduction of a larger system. Also, the ideas of discrete reduction are very useful in the block diagonalization stability analysis of the energy momentum method, even though the reduction is not carried out explicitly.

To help understand the reduction of P_Σ, we consider some things that are relevant for the momentum map of the action of G_Σ. The following prepatory lemma is standard (see, for example, Guillemin and Prato [1990]) but we shall need the proof for later developments, so we give it.

Lemma 8.1 *The Lie algebra $\mathfrak{g}$ of G splits as follows:*

$$\mathfrak{g} = \mathfrak{g}_\Sigma \oplus (\mathfrak{g}_\Sigma^*)^0 \qquad (8.1.10)$$

where $\mathfrak{g}_\Sigma^* = \mathrm{Fix}(\Sigma, \mathfrak{g}^*)$ is the fixed point set for the action of Σ on $\mathfrak{g}^*$ and where the superscript zero denotes the annihilator in $\mathfrak{g}$. Similarly, the dual splits as

$$\mathfrak{g}^* = \mathfrak{g}_\Sigma^* \oplus (\mathfrak{g}_\Sigma)^0. \qquad (8.1.11)$$

Proof First, suppose that $\xi \in \mathfrak{g}_\Sigma \cap (\mathfrak{g}_\Sigma^*)^0$, $\mu \in \mathfrak{g}^*$ and $\sigma \in \Sigma$. Since $\xi \in \mathfrak{g}_\Sigma$,

$$\langle \mu, \xi \rangle = \langle \mu, \sigma_\mathfrak{g} \cdot \xi \rangle = \langle \sigma_{\mathfrak{g}^*}^{-1} \cdot \mu, \xi \rangle. \qquad (8.1.12)$$

Averaging (8.1.12) over σ relative to an invariant measure on Σ, which is possible since Σ is compact, gives: $\langle \mu, \xi \rangle = \langle \bar{\mu}, \xi \rangle$. Since the average $\bar{\mu}$ of μ is Σ-invariant, we have $\bar{\mu} \in \mathfrak{g}_\Sigma^*$ and since $\xi \in (\mathfrak{g}_\Sigma^*)^0$ we get

$$\langle \mu, \xi \rangle = \langle \bar{\mu}, \xi \rangle = 0.$$

Since μ was arbitrary, $\xi = 0$. Thus, $\mathfrak{g}_\Sigma \cap (\mathfrak{g}_\Sigma^*)^0 = \{0\}$. Using an invariant metric on $\mathfrak{g}$, we see that $\dim \mathfrak{g}_\Sigma = \dim(\mathfrak{g}_\Sigma^*)^0$ and thus $\dim \mathfrak{g} = \dim \mathfrak{g}_\Sigma + \dim(\mathfrak{g}_\Sigma^*)^0$, so we get the result (8.1.10). The proof of (8.1.11) is similar. (If the group is infinite dimensional, then one needs to show that every element can be split; in practice, this usually relies on a Fredholm alternative-elliptic equation argument.) ∎

It will be useful later to note that this argument also proves a more general result as follows:

Lemma 8.2 *Let Σ act linearly on a vector space W. Then*

$$W = W_\Sigma \oplus (W_\Sigma^*)^0 \qquad (8.1.13)$$

where W_Σ is the fixed point set for the action of Σ on W and where $(W_\Sigma^*)^0$ is the annihilator of the fixed point set for the dual action of Σ on W^*. Similarly, the dual splits as

$$W^* = W_\Sigma^* \oplus (W_\Sigma)^0. \qquad (8.1.14)$$

Adding these two splittings gives the splitting

$$W \times W^* = (W_\Sigma \times W_\Sigma^*) \oplus ((W_\Sigma^*)^0 \times (W_\Sigma)^0) \qquad (8.1.15)$$

which is the splitting of the symplectic vector space $W \times W^*$ into the symplectic subspace $W_\Sigma \times W_\Sigma^*$ and its symplectic orthogonal complement. The splitting (8.1.10) gives a natural identification

$$(\mathfrak{g}_\Sigma)^* \cong \mathfrak{g}_\Sigma^*.$$

Note that the splittings (8.1.10) and (8.1.11) do not involve the choice of a metric; that is, the splittings are natural.

The following **block diagonalization into isotypic components** will be used to prove the subblocking theorem in the energy-momentum method.

Lemma 8.3 *Let $B : W \times W \to \mathbb{R}$ be a bilinear form invariant under the action of Σ. Then the matrix of B block diagonalizes under the splitting (8.1.13). That is,*

$$B(w, u) = 0 \quad and \quad B(u, w) = 0$$

if $w \in W_\Sigma$ and $u \in (W_\Sigma^*)^0$.

Proof Consider the element $\alpha \in W^*$ defined by $\alpha(v) = B(w, v)$. It suffices to show that α is fixed by the action of Σ on W^* since it annihilates W_Σ^*. To see this, note that invariance of B means $B(\sigma w, \sigma v) = B(w, v)$, *i.e.*, $B(w, \sigma v) = B(\sigma^{-1}w, v)$. Then

$$(\sigma^* \alpha)(v) = \alpha(\sigma v) = B(w, \sigma v) = B(\sigma^{-1}w, v) = B(w, v) = \alpha(v)$$

since $w \in W_\Sigma$ so w is fixed by σ. This shows that $\alpha \in W_\Sigma^*$ and so $\alpha(u) = B(w, u) = 0$ for u in the annihilator. The proof that $B(u, w) = 0$ is similar. ∎

For the classical water molecule, W_Σ corresponds to *symmetric variations* of the molecule's configuration and $(W_\Sigma^*)^0$ to *non-symmetric* ones.

Proposition 8.3 $\mathbf{J}(P_\Sigma) \subset \mathfrak{g}_\Sigma^*$ *and the momentum map for the* G_Σ *action on* P_Σ *is* $\mathbf{J}$ *restricted to* P_Σ, *taking values in* $\mathfrak{g}_\Sigma^* \cong (\mathfrak{g}_\Sigma)^*$.

Proof Let $\sigma \in \Sigma$ and $z \in P_\Sigma$. By (8.1.5),

$$\mathfrak{g}_{\mathfrak{g}^*} \cdot \mathbf{J}(z) = \mathbf{J}(\sigma_P(z)) = \mathbf{J}(z),$$

so $\mathbf{J}(z) \in \mathrm{Fix}(\Sigma, \mathfrak{g}^*)$. The second result follows since the momentum map for G_Σ acting on P_Σ is $\mathbf{J}$ composed with the projection of $\mathfrak{g}^*$ to $\mathfrak{g}_\Sigma^*$. ∎

Next, we consider an interesting transversality property of the fixed point set. Assume that each $\mu \in \mathfrak{g}_\Sigma^*$ is a regular value of $\mathbf{J}$. Then

$$W_\Sigma = \mathbf{J}^{-1}(\mathfrak{g}_\Sigma^*) \subset P$$

is a submanifold, $P_\Sigma \subset W_\Sigma$, and $\pi = \mathbf{J}|W_\Sigma : W_\Sigma \to \mathfrak{g}_\Sigma$ is a submersion.

Proposition 8.4 *The manifold* P_Σ *is transverse to the fibers* $\mathbf{J}^{-1}(\mu) \cap W_\Sigma$.

Proof Recall that P_Σ is a manifold with tangent space at $z \in P_\Sigma$ given by

$$T_z P_\Sigma = \{v \in T_z P \,|\, T_z \sigma_P \cdot b = v \quad \text{for all} \quad \sigma \in \Sigma\}.$$

Choose $\nu \in \mathfrak{g}_\Sigma^*$ and $u \in T_z W_\Sigma$ such that $T\pi \cdot u = \nu$. Note that if $\sigma \in \Sigma$, then $T\pi \cdot T\sigma \cdot u = \sigma_{\mathfrak{g}^*} T\pi \cdot u = \sigma_{\mathfrak{g}^*} \cdot \nu = \nu$, so $T\sigma \cdot u$ has the same projection as u. Thus, if we average over Σ, we get $\bar{u} \in T_z W_\Sigma$ also with $T\pi \cdot \bar{u} = \nu$. But since $\bar{u}$ is the average, it is fixed by each $T_z \sigma$, so $\bar{u} \in T_z P_\Sigma$. Thus, $T_z P_\Sigma$ is transverse to the fibers of π. ∎

See Sjamaar [1990], Lemma 3.2.3 for a related result. In particular, this result implies that

Corollary 8.1 *The set of* $\nu \in \mathfrak{g}_\Sigma^*$ *for which there is a* Σ-*fixed point in* $\mathbf{J}^{-1}(\nu)$, *is open.*

8.2 Cotangent Bundles

Let $P = T^*Q$ and assume that Σ and G act on Q and hence on T^*Q by cotangent lift. Assume that Σ acts on G but now assume the actions are compatible in the following sense:

Assumption 1_Q. *The following equation holds* (Figure 8.2.1):

$$\sigma_Q \circ g_Q = [\sigma_G(g)]_G \circ \sigma_Q. \tag{8.2.1}$$

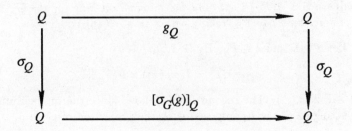

Figure 8.2.1: Assumption 1_Q.

Proposition 8.5 *Under Assumption* 1_Q, *both Assumptions 1 and 2 are valid.*

Proof Equation (8.1.1) follows from (8.2.1) and the fact that cotangent lift preserves compositions. Differentiation of (8.2.1) with respect to g at the identity in the direction $\xi \in \mathfrak{g}$ gives

$$T\sigma_Q \circ \xi_Q = (\sigma_{\mathfrak{g}} \cdot \xi)_Q \circ \sigma_Q. \qquad (8.2.2)$$

Evaluating this at q, pairing the result with a covector $p \in T^*_{\sigma_Q(q)}Q$, and using the formula for the momentum map of a cotangent lift gives

$$\langle p, T\sigma_Q \cdot \xi_Q(q) \rangle = \langle p, (\sigma_{\mathfrak{g}} \cdot \xi)_Q(\sigma_Q(q)) \rangle$$

i.e.,

$$\langle T\sigma_Q^* \cdot p, \xi_Q(q) \rangle = \langle p, (\sigma_{\mathfrak{g}} \cdot \xi)_Q(\sigma_Q(q)) \rangle$$

i.e.,

$$\langle \mathbf{J}(T\sigma_Q^* \cdot p), \xi \rangle = \langle \mathbf{J}(p), \sigma_{\mathfrak{g}} \cdot \xi \rangle = \langle \sigma_{\mathfrak{g}^*}^{-1} \mathbf{J}(p), \xi \rangle$$

i.e.,

$$\mathbf{J} \circ \sigma_P^{-1} = \sigma_{\mathfrak{g}^*}^{-1} \circ \mathbf{J}$$

so (8.1.5) holds. ∎

Proposition 8.6 *For cotangent lifts, and* $Q_\Sigma = \text{Fix}(\Sigma, Q)$, *we have*

$$P_\Sigma = T^*(Q_\Sigma) \qquad (8.2.3)$$

with the canonical cotangent structure. Moreover, the action of G_Σ *is the cotangent lift of its action on* Q_Σ *and its momentum map is the standard one for cotangent lifts.*

For (8.2.3) to make sense, we need to know how $T_q^* Q_\Sigma$ is identified with a subspace of $T_q^* Q$. To see this, consider the action of Σ on $T_q^* Q$ by

$$\sigma \cdot \alpha_q = \left([T\sigma_Q(q)]^{-1}\right)^* \cdot \alpha_q$$

and regard $\text{Fix}(\Sigma, T_q^* Q)$ as a linear subspace of $T_q^* Q$.

Lemma 8.4

$$T_q Q = T_q Q_\Sigma \oplus [\text{Fix}(\Sigma, T_q^* Q)]^0. \tag{8.2.4}$$

Proof As before, $T_q Q_\Sigma = \text{Fix}(\Sigma, T_q Q)$. We prove the lemma by a procedure similar to Lemma 8.1. Let $v \in T_q Q_\Sigma \cap \text{Fix}(\Sigma, T_q^* Q)^0, \xi \in T_q^* Q$, and $\sigma \in \Sigma$. Then

$$\langle \alpha, v \rangle = \langle \alpha, T\sigma_Q \cdot v \rangle = \langle T\sigma_Q^{-1} \cdot \alpha, v \rangle. \tag{8.2.5}$$

Averaging (8.2.5) over σ gives

$$\langle \sigma, v \rangle = \langle \bar{\alpha}, v \rangle = 0$$

so $v = 0$. The result follows by a dimension count, as in Lemma 8.1. ∎

If $\alpha_q \in T_q^* Q_\Sigma$, extend it to $T_q^* Q$ by letting it be zero on $[\text{Fix}(\Sigma, T_q^* Q)]^0$. This embeds $T_q^* Q_\Sigma$ into $T_q^* Q$ and provides the split

$$T_q^* Q = T_q^* Q_\Sigma \oplus (T_q Q_\Sigma)^0 \tag{8.2.6}$$

identifying $T_q^* Q_\Sigma \cong \text{Fix}(\Sigma, T_q^* Q)$.

Returning to the proof of Proposition 8.5, by definition of $P_\Sigma, \alpha_q \in P_\Sigma$ iff $\sigma_P(\alpha_q) = \alpha_q$ for all $\sigma \in \Sigma$; *i.e.*,

$$\left([T\sigma_Q(q)]^{-1}\right)^* \alpha_q = \alpha_q$$

i.e., $\alpha_q \in \text{Fix}(\Sigma, T_q^* Q) = T^* Q_\Sigma$. Thus, $P_\Sigma = T^*(Q_\Sigma)$. The symplectic structure on P_Σ is obtained from $T^* Q$ by restriction. We need to show that the inclusion map $i_\Sigma : T^* Q_\Sigma \to T^* Q$ defined by (8.2.6) is a symplectic embedding. In fact, it is readily checked that i_Σ pulls the canonical one form on $T^* Q$ back to that on $T^* Q_\Sigma$ because the projection $\pi_\Sigma : T^* Q \to P$ satisfies $\pi \circ i_\Sigma = \pi_\Sigma$. The rest of the proposition now follows. ∎

8.3 Examples

Example 1 Let the phase space be $P = \mathbb{C}^2$ with the symplectic structure given by

$$\Omega((z_1, w_1), (z_2, w_2)) = 2\text{Im}(z_1 \bar{z}_2 + w_1 \bar{w}_2).$$

Let $G = U(2)$ act simply by matrix multiplication and let $\Sigma = \mathbb{Z}_2$ act on P by $\sigma(z, w) = (w, z)$, where σ denotes the nontrivial element of Σ. Identifying σ with the matrix $\begin{bmatrix} 0 & 1 \\ 1 & 0 \end{bmatrix}$, the actions are compatible (Assumption 1) if we let Σ act on G by conjugation: $\sigma \cdot A = \sigma A \sigma^{-1}$.

Note that the actions of G and Σ on P are symplectic. The Lie algebra $u(2)$ of $U(2)$ is identified with the skew hermetian matrices, which we write as

$$\xi = \begin{bmatrix} iu & \beta \\ -\bar{\beta} & iv \end{bmatrix}$$

where u and v are real and $\beta \in \mathbb{C}$. The equivariant momentum map for the $U(2)$ action is given by

$$\langle \mathbf{J}(z, w), \xi \rangle = \frac{1}{2} \Omega \left(\xi \begin{bmatrix} z \\ w \end{bmatrix}, \begin{bmatrix} z \\ w \end{bmatrix} \right) = u|z|^2 + v|w|^2 + 2\mathrm{Im}(\beta \bar{z} w).$$

Assumption 2 reads

$$\langle \mathbf{J}(\sigma(z, w)), \xi \rangle = \langle \mathbf{J}(z, w), \sigma \xi \sigma^{-1} \rangle,$$

which is easily checked.

In this example, G_Σ consists of elements of G that commute with σ; that is, G_Σ is the subgroup of $U(2)$ consisting of matrices of the form $\begin{bmatrix} a & b \\ b & a \end{bmatrix}$. Note that $P_\Sigma = \{(z, z) \in P \mid z \in \mathbb{C}\}$ and that the G_Σ action on P_Σ is $(z, z) \mapsto ((a + b)z, (a + b)z)$. We identify the Lie algebra of G_Σ with $\mathfrak{g}_\Sigma$ consisting of matrices of the form $\xi = i \begin{bmatrix} u & y \\ y & u \end{bmatrix}$ where u and y are real. Note that

$$\langle \mathbf{J}(z, z), \xi \rangle = 2(u + y)|z|^2,$$

consistent with Proposition 8.3. In §4, we shall see an alternative way of viewing this example in terms of dual pairs using $SU(2)$ and S^1 separately, rather than as $U(2)$. This will also bring out links with the one-to-one resonance more clearly. ♦

The next example is an elementary physical example in which we deal with cotangent bundles.

Example 2 We consider the spherical pendulum with $P = T^*Q$ where $Q = S_l^2$, the two sphere of radius l. Here $G = S^1$ acts on Q by rotations about the z-axis so that the corresponding momentum map is simply the angular momentum about the z-axis (with the standard identifications). We let $\Sigma = \mathbb{Z}_2$ act on S_l^2 by reflection in a chosen plane, say the yz-plane, so $\sigma(z, y, z) = (-x, y, z)$, where $q = (x, y, z) \in Q$. This action is lifted to P by cotangent lift, so it is *symplectic*. We let σ act on G by conjugation; if R_θ is the rotation about the z-axis through an angle θ, then

$$\sigma \cdot R_\theta = \sigma R_\theta \sigma^{-1} = R_{-\theta}.$$

Note that the group $\Sigma \circledS G$ is $O(2)$. To check Assumptions 1 and 2, it suffices to check Assumption 1_Q. It states that

$$\sigma(R_\theta \cdot q) = (\sigma R_\theta \sigma^{-1})(\sigma(q))$$

which is obviously correct. Here $P_\Sigma = T^*(Q_\Sigma)$ where $Q_\Sigma = S_l^1$ is S_l^2 intersect the yz-plane. Restriction to $T^*(Q_\Sigma)$ gives a simple pendulum moving in the yz-plane. Here $G_\Sigma = \{e, R_\pi\}$, so the discrete reduced space is $P_\Sigma/\mathbb{Z}_2$, a symplectic orbifold (see Sjamaar [1990]). Thus, we essentially find that *discrete reduction of the spherical pendulum is the simple pendulum moving in the yz-plane*. Note that here one has different choices of Σ that correspond to different planes of swing of the simple pendulum. Note also, consistent with this, that the angular momentum vanishes on P_Σ. ♦

Example 3 Here we consider the *double spherical pendulum*. Let $P = T^*Q$ where $Q = S_{l_1}^2 \times S_{l_2}^2$ and $(\mathbf{q}_1, \mathbf{q}_2) \in Q$ gives the configuration of the two arms of the pendulum. Again, we let $G = S^1$ consist of rotations about the z-axis, with G acting by the diagonal action on Q. We again let $\Sigma = \mathbb{Z}_2$, and let σ act by reflection in the yz-plane, acting simultaneously on both factors in $Q = S^2 \times S^2$. As in Example 2, *this leads via discrete reduction from the spherical double pendulum to the planar double pendulum* and $G_\Sigma = \{e, R_\pi\}$.

Another way $\mathbb{Z}_2$ acts on Q that is useful in our study of relative equilibria is as follows.

Let $\sigma \in \mathbb{Z}_2$ be the nontrivial element of $\mathbb{Z}_2$ and let σ map $(\mathbf{q}_1, \mathbf{q}_2)$ to $(\mathbf{q}_1, \sigma_p \mathbf{q}_2)$, where P is the vertical plane spanned by $\mathbf{k}$ and q_1 and where σ_P is the reflection in this plane. This time, Σ acts *trivially* on G, so $\Sigma \circledS G$ is the *direct product*. Compatible with the general theory in the next section, the subblocking property associated with this symmetry is what gave the subblocking property we observed in Equation (5.5.4) in which we calculated $\delta^2 V_\mu$ for purposes of the stability analysis.

Notice that interchanging the two pendula does *not* produce a discrete symmetry because the presense of gravity leads to a noninvariance of the kinetic and potential energies under particle interchange.   ♦

An example with a slightly richer discrete symmetry group is the system of *two whirling masses*.

Example 4 Here we let $P = T^*Q$ where $Q = \mathbb{R}^3 \times \mathbb{R}^3$ and $(\mathbf{q}_1, \mathbf{q}_2) \in Q$ give the configuration of the two masses of the system, as in Figure 8.3.1. The masses are connected to each other and to the supports by springs. The Lagrangian is the standard one: kinetic minus potential energy. However, in this example, the potential energy comes only from the springs; gravity is ignored.

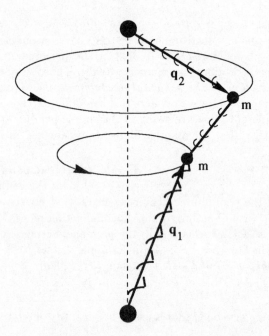

Figure 8.3.1: Two whirling masses.

Let $G = S^1$ act by rotations of both masses about the (vertical) z-axis. The momentum map is the angular momentum about the vertical axis. We again let $\Sigma = \mathbb{Z}_2$, but now there are two cases:
Case 1 Here let σ act by reflection in a *vertical* plane, say the yz-plane, acting simultaneously on both factors in Q. As above, this action produces an associated invariant subsystem on its fixed point space, the problem

of two *planar* masses connected by springs. This is the case even if the masses of the two particles are different and if the springs are different. As in Example 2, the group $\Sigma \circledS G$ is $O(2)$. As with the double spherical pendulum we can consider the discrete symmetry of reflection in a moving vertical plane. This is important again in the study of relative equilibria.
Case 2 Here we let the masses of the two particles be the same, the two outside springs be identical and all three springs be reflection invariant. Now let Σ act on Q by simultaneous reflection in the horizontal plane (and translation of the vector base points to the opposite support). In this case the actions of G and S on Q (and hence on P) commute, so we can let Σ act trivially on G and Assumption 1_Q holds. Here, $\Sigma \circledS G = \mathbb{Z}_2 \times SO(2)$, a *direct* product. Now $G_\Sigma = S^1$ and $P_\Sigma = T^* \mathbb{R}^3$ corresponds to identical motions of the two masses. Thus, *in this case discrete reduction gives the motion of a symmetric two mass system on which S^1 still acts.*

This richer collection of discrete symmetries then naturally leads to a rich block structure in the second variation $\delta^2 V_\mu$ and a simpler normal form for the linearized equations at a relative equilibrium. See Zombro and Holmes [1991] for more information. ◆

Example 5 The rigid body has three copies of $\mathbb{Z}_2$ as symmetry group. Here $Q = SO(3)$ and we choose a plane P corresponding to one of the principle moments of inertia and let σ be the reflection in this plane. Let $\Sigma = \mathbb{Z}_2$ be generated by σ. Let $P = T^*Q$, and $G = SO(3)$. Let G act on Q by left multiplication (as is usual for the rigid body), so the momentum map is the spatial angular momentum of the body. The discrete symmetries in this example yield the well known symmetry of the phase portrait of the rigid body as viewed on the two sphere. Let Σ act on Q by conjugation; for $A \in Q, \sigma \cdot A = \sigma A \sigma^{-1}$. If we let Σ act on G the same way, then

$$\sigma_Q(R_Q(A)) = \sigma R A \sigma^{-1} = (\sigma R \sigma^{-1})(\sigma A \sigma^{-1}) = \sigma_G(R) \cdot \sigma_Q(A),$$

so Assumption 1_Q holds. As in Example 2, the semidirect product is the orthogonal group: $\Sigma \circledS G = O(3)$. In this case, G_Σ consists of rotations in the plane defining σ and $P_\Sigma = T^* G_\Sigma$. Here, *the discrete reduction yields a rigid rotor constrained to rotate about a fixed principal axis.*

Remark This example illustrates two cautions that are needed when dealing with discrete symplectic symmetries. First, while there is a well defined action of Σ on P/G, this action need not be Poisson. For example the induced action on $so(3)^*$ in Example 5 is anti-Poisson. Also, for $\sigma \in \Sigma, \sigma$ maps $\mathbf{J}^{-1}(\mu)$ to $\mathbf{J}^{-1}(\sigma_{\mathfrak{g}^*} \mu)$, so there need not be a well defined action on P_μ, let alone a symplectic one. (For the rigid body one gets an *anti-symplectic* map of S^2 to itself.)   ♦

Example 6 We modify Example 5 to a situation that is of interest
in pseudo rigid bodies and gravitating fluid masses (see Lewis and Simo
[1990]). Let $Q = GL(3)$ (representing linear, but nonrigid deformations of
a reference configuration). Let $G = SO(3)$ act on the left on Q as before, so
again, the momentum map represents the total spatial angular momentum
of the system. As in Example 5, let $\Sigma = \mathbb{Z}_2$ and let σ be reflection in a
chosen, fixed plane P. Let σ_S denote reflection in the image of the plane P
under linear transformation S, and let the action of σ on an element A of
Q be on the right by σ and on the left by σ_A and let it act trivially on G.
Again, one checks that Assumption 1_Q holds. In this case, the semidirect
product is the direct product $\mathbb{Z}_2 \times SO(3)$, so the full rotation group still
acts on $P_\Sigma =$. Now, Q_Σ represents configurations that have a plane of
symmetry, so in this case, *discrete reduction of this system correponds to
restriction to symmetric bodies.*

In view of the block diagonalization results outlined in the next section,
it is of interest to not to actually carry out the discrete reduction, but rather
to use the discrete symmetry to study stability. This seems to explain, in
part, why, in their calculations, further subblocking, in some cases, even to
diagonal matrices, was found.   ♦

8.4   Sub-Block Diagonalization with Discrete Symmetry

In this section, we explain how Lemma 8.3 gives a subblocking theorem
in the energy-momentum method. We indicate how this result applies to
the classical water molecule without carrying out the calculations in detail
here. In this procedure, we do not carry out discrete reduction explicitly
(although it does give an interesting invariant subsystem for the dynamics of
symmetric molecules). Rather, we use it to divide the modes into symmetric
and non-symmetric ones.

Recall from Chapter 5, that one splits the space of variations of the
concrete realization $\mathcal{V}$ of Q/G_μ into variations $\mathcal{V}_{\mathrm{RIG}}$ in G/G_μ and variations
$\mathcal{V}_{\mathrm{INT}}$ in Q/G_μ. With the appropriate splitting, one gets the block diagonal
structure

$$\delta^2 V_\mu = \begin{bmatrix} \text{Arnold form} & 0 \\ 0 & \text{Smale form} \end{bmatrix}$$

where the Arnold form means $\delta^2 V_\mu$ computed on the coadjoint orbit tangent space, and the Smale form means $\delta^2 V_\mu$ computed on Q/G. Perhaps even more interesting is the structure of the linearized *dynamics* near a relative equilibrium. That is, both the augmented Hamiltonian $H_\xi = H - \langle \mathbf{J}, \xi \rangle$ *and* the symplectic structure can be *simultaneously* brought into the following normal form:

$$
\delta^2 H_\xi = \begin{bmatrix} \text{Arnold form} & 0 & 0 \\ 0 & \text{Smale form} & 0 \\ 0 & 0 & \text{Kinetic Energy} > 0 \end{bmatrix}
$$

and

$$
\text{Symplectic Form} = \begin{bmatrix} \text{coadjoint orbit form} & C & 0 \\ -C & \text{magnetic (coriolis)} & I \\ 0 & -I & 0 \end{bmatrix}
$$

where the columns represent the **coadjoint orbit variables** (G/G_μ), the **shape variables** (Q/G) and the **shape momenta** respectively. The term C is an **interaction term** between the group variables and the shape variables. The magnetic term is the curvature of the μ-component of the mechanical connection, as we described earlier.

Suppose that we have a compact discrete group Σ acting by isometries on Q, and preserving the potential, so we are in the setting of the preceeding section. This action lifts to the cotangent bundle, as we have seen. The resulting fixed point space is the cotangent bundle of the fixed point space Q_Σ. This fixed point space represents the Σ-symmetric configurations.

Of more concern here is the fact that the action also gives an action on the quotient space, or shape space Q/G. We can split the tangent space to Q/G at a configuration corresponding to the relative equilibrium according to Lemma 8.2 and can apply Lemma 8.3 to the Smale form. Here, one must check that the amended potential is actually invariant under Σ. In general, this need not be the case, since the discrete group need not leave the value of the momentum μ invariant. However, there are two important cases for which this is verified. The first is for $SO(3)$ with $\mathbb{Z}_2$ acting by conjugation, as in the rigid body example, it maps μ to its negative, so in this case, from the formula $V_\mu(q) = V(q) - \mu \mathbb{I}(q)^{-1} \mu$ we see that indeed V_μ is invarint. The second case, which is relevant for the water molecule, is when Σ acts trivially on G. Then Σ leaves μ invariant, and so V_μ is again invariant.

Theorem 8.1 *Under these assumptions, at a relative equilibrium, the second variation $\delta^2 V(q)$ block diagonalizes, which we refer to as the **subblocking** property.*

The blocks in the Smale form are given by Lemma 8.2: they are the Σ-symmetric variations, and their complement chosen according to that lemma as the annihilator of the symmetric dual variations. Lemma 8.3 can also be applied to the symplectic form, showing that it subblocks as well.

These remarks apply to the classical rotating water molecule as follows. (We let the reader work out the case of the two whirling masses in a similar, but simpler vein.) The discrete symmetry Σ is $\mathbb{Z}_2$ acting, roughly speaking, by interchanging the two hydrogen atoms. We shall describe the action more precisely in a moment. The action of $\mathbb{Z}_2$ on $SO(3)$ will be the trivial one, so the semi-direct product is the direct product and so G_Σ is again $SO(3)$. The discrete reduction procedure then yields the system consisting of symmetric molecules (with the two hydrogen atoms moving in a symmetric way), but still with symmetry group the rotation group, which makes good sense physically. The subblocking property mentioned above shows that the Smale form, which in this example, is a 3×3 symmetric matrix, becomes block diagonal, with a 2×2 subblock corresponding to the symmetric variations, and a singleton block corresponding to the nonsymmetric variations.

Explicitly, the nontrivial element σ of Σ acts on the configuration space as follows:

$$\sigma(\mathbf{r}, \mathbf{s}) = \left(\mathbf{r}, \mathbf{s} - 2(\mathbf{r} \cdot \mathbf{s}) \frac{\mathbf{r}}{\| \mathbf{r} \|^2} \right).$$

This symmetry interchanges the role of the two masses as in Figure 8.4.1.

If one prefers, σ is the transformation that rotates the system by 180 degrees about a line perpendicular to $\mathbf{r}$ in the plane of the system, followed by an interchange of the particles. One now checks the claims made earlier; for example, that the Assumption 1_Q is satisfied, and that the fixed point set of the discrete symmetry group does correspond to the symmetric configurations, etc. Notice that for the ozone molecule, one can apply this operation to any two pairs of atoms, and this leads to interesting consequences (such as, for the planar molecule, the "breathing mode" decouples dynamically from the other modes, and in space, it couples very simply via the internal rigid coupling discussed in Chapter 5.

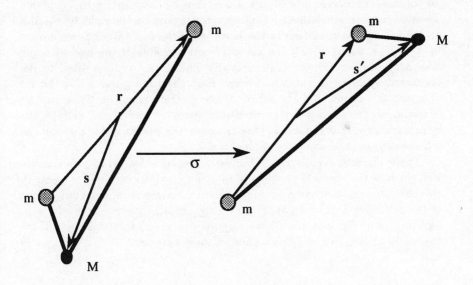

Figure 8.4.1: Discrete symmetry of the water molecule.

We conclude that $\delta^2 V_\mu$ takes the following form:

$$\delta^2 V_\mu = \begin{bmatrix} a & b & 0 \\ b & d & 0 \\ 0 & 0 & g \end{bmatrix} \tag{8.4.1}$$

where the block $\begin{bmatrix} a & b \\ b & d \end{bmatrix}$ corresponds to the symmetric variations and $[g]$ to the symmetry breaking variations.

Another system in which an interesting discrete symmetry occurs is the rotating liquid drop, as studied by Lewis, Marsden and Ratiu [1987] and Lewis [1989]. Here, one has an ideal incompressible, inviscid fluid system (so the system is infinite dimensional) that has a basic symmetry group S^1 for a planar drop, and $SO(3)$ for a drop in space. In either case, the corresponding momentum map is the angular momentum of the drop. However, there is a discrete symmetry as well, that can be set up in a way somewhat similar to that of the rigid body, described in §8.3. For the planar case, one chooses a line say $\mathcal{L}$ in the reference configuration of the fluid and calls the reflection in $\mathcal{L}$ by $\sigma_{\mathcal{L}}$. The configuration space for the drop is $Q =$

all volume preserving embeddings of a reference configuration to the plane. Then $\sigma_{\mathcal{L}}$ acts on an element $\eta \in Q$ by conjugation, on the right by $\sigma_{\mathcal{L}}$ and on the left by the reflection in the line that is the line through the images $\eta(P_1), \eta(P_2)$, where P_1, P_2 are the intersection points of the line with the boundary of the reference configuration. This action is then lifted to the cotangent space in the standard way. Here the fixed point set is the set of symmetric drops, and the action of the discrete group on the group S^1 is trivial, so the fixed point group G_Σ is again S^1 and it still acts on the symmetric drops, as it should. One can view the planar water molecule as a finite dimensional analogue of this model.

These discrete symmetries are clearly available for other problems as well, such as the classical problem of rotating gravitational fluid masses. In this case, one has a richer symmetry structure coming from finite subgroups of the orthogonal group $O(3)$. From the general discussion above, the sub-blocking property that the discrete symmetry gives, should be useful for the study of stability and bifurcation of these systems.

8.5   Discrete Reduction of Dual Pairs

We now give a set up one can use in the application of the discrete reduction procedure to integrable Hamiltonian systems. Consider two Lie groups G and H acting symplectically on the manifold P, and assume that the actions commute. If $\mathfrak{g}$ and $\mathfrak{h}$ denote the Lie algebras of G and H respectively, we suppose that there are equivariant momentum maps

$$\mathbf{J}_G : P \to \mathfrak{g}^* \quad and \quad \mathbf{J}_H : P \to \mathfrak{h}^* \tag{8.5.1}$$

generating the actions.

Dual Pair Assumption *For all $z \in P$:*

$$\mathbf{J}_G^{-1}(\mathbf{J}_G(z)) = \mathcal{O}_z^H \quad and \quad \mathbf{J}_H^{-1}(\mathbf{J}_H(z)) = \mathcal{O}_z^G \tag{8.5.2}$$

where $\mathcal{O}_z^H$ and $\mathcal{O}_z^G$ denote the orbits of H and G respectively through z. We also assume that the group Σ acts on $P, G,$ and H and hence on $\mathfrak{g}, \mathfrak{h}$ and $\mathfrak{g}^*$ and $\mathfrak{h}^*$ where Assumptions 1 and 2 of §8.1 hold for both G and H (Figures 8.5.1 and 8.5.2).

Assuming also that the action of H is free and proper, so that

$$P \to P/H$$

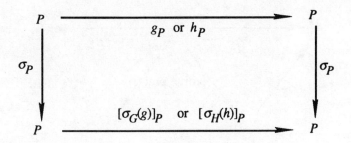

Figure 8.5.1: Assumption 1 for dual pairs.

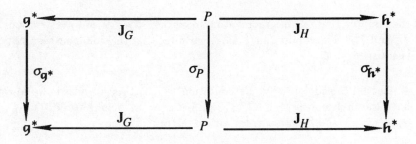

Figure 8.5.2: Assumption 2 for dual pairs.

is an H-fibration, it follows that the mapping induced by $\mathbf{J}$

$$\overline{\mathbf{J}}_G : P/H \to \mathfrak{g}^*$$

is a Poisson embedding, where $\mathfrak{g}^*$ has the Lie-Poisson structure (the minus structure if the actions are left, and the plus structure if the actions are on the right) and where P/H has the quotient Poisson structure. Thus, we have the commutative diagram in Figure 8.5.3.

For the following proposition, note that $\mathcal{O}_z^{G_\Sigma} \subset \mathcal{O}_z^G \cap P_\Sigma$ and similarly for H.

Proposition 8.7 *Let (P, G, H) be a dual pair, acted on, as above, by Σ. Suppose that for each $z \in P_\Sigma$,*

$$\mathcal{O}_z^{G_\Sigma} = \mathcal{O}_z^G \cap P_\Sigma \quad and \quad \mathcal{O}_z^{H_\Sigma} = \mathcal{O}_z^H \cap P_\Sigma. \qquad (8.5.3)$$

Then $(P_\Sigma, G_\Sigma, H_\Sigma)$ form a dual pair.

Proof That P_Σ is a symplectic manifold, with commuting Hamiltonian actions of G_Σ, H_Σ follows from Propositions 8.1 and 8.2. Proposition 8.3

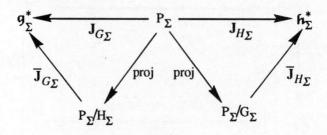

Figure 8.5.3: Reduction of dual pairs.

then implies that one has momentum maps $\mathbf{J}_{G_\Sigma}, \mathbf{J}_{H_\Sigma}$ given by restrictions of $\mathbf{J}_G$ and $\mathbf{J}_H$. The conditions (8.5.3) guarantee that the dual pair assumption is satisfied. ∎

Let us consider briefly the condition (8.5.3). As we noted, we always have the inclusion $\mathcal{O}_z^{H_\Sigma} \subset \mathcal{O}_z^H \cap P_\Sigma$. Let $z' = hz$ belong to $\mathcal{O}_z^H \cap P_\Sigma$. As $\sigma_P(z') = z'$, and $\sigma_P(z) = z$, for all $\sigma \in \Sigma$, one obtains:

$$h^{-1}\sigma_H(h) \in \operatorname{Stab}_H(z).$$

If σ is an involution, $h^{-1}\sigma_H(h)$ belongs to the subset $\operatorname{Stab}_H^-(z)$ of elements t satisfying $\sigma(t) = t^{-1}$. One wants an $\hat{h} = hs^{-1}$, where $s \in \operatorname{Stab}_H(z)$, such that $\sigma_H(\hat{h}) = \hat{h}$, for all $\sigma \in \Sigma$. One must then solve the equations:

$$h^{-1}\sigma(h) = s^{-1}\sigma(s) \quad \text{for all} \quad \sigma \in \Sigma$$

for some $s \in \operatorname{Stab}_H(z)$. This is frequently possible: *one simple case is when the action of H is free.* When Σ is generated by an involution σ, (*i.e.*, $\Sigma = \mathbb{Z}_2$), the preceding equation will be solvable if the map

$$\Pi_\sigma : \operatorname{Stab}_H(z) \to \operatorname{Stab}_H^-(z); s \mapsto s^{-1}\sigma(s)$$

is surjective.

If the action of H (and so of H_Σ) is free and proper and $\mathbf{J}_G$ is a submersion along $\mathbf{J}_G^{-1}(\mathfrak{g}_\Sigma^*)$ (or at least over some open subset of $\mathfrak{g}_\Sigma^*$), then one has an H-bundle in P

$$\mathbf{J}_G : \mathbf{J}_G^{-1}(\mathfrak{g}_\Sigma^*) \to \mathbf{J}_G^{-1}(\mathfrak{g}_\Sigma^*)/H \subset \mathfrak{g}_\Sigma^* \tag{8.5.4}$$

and an H_Σ bundle in P_Σ

$$\mathbf{J}_{G_\Sigma} : \mathbf{J}_{G_\Sigma}^{-1} : (\mathfrak{g}_\Sigma^*) \to \mathbf{J}_{G_\Sigma}^{-1}(\mathfrak{g}_\Sigma^*)/H_\Sigma \subset \mathfrak{g}_\Sigma^* \tag{8.5.5}$$

giving an inclusion:

$$\mathbf{J}_{G_\Sigma}^{-1}(\mathfrak{g}_\Sigma^*)/H_\Sigma \subset \mathbf{J}_G^{-1}(\mathfrak{g}_\Sigma^*)/H.$$

Over $\mathbf{J}_{G_\Sigma}^{-1}(\mathfrak{g}_\Sigma^*)/H_\Sigma$, (8.5.5) is a subbundle of (8.5.4). This, in fact, occurs over an open set:

Proposition 8.8 *Under the hypotheses given above,* $\mathbf{J}_{G_\Sigma}^{-1}(\mathfrak{g}_\Sigma^*)/H_\Sigma$ *is an open subset of* $\mathbf{J}_G^{-1}(\mathfrak{g}_\Sigma^*)/H$.

Proof $\mathbf{J}_{G_\Sigma}^{-1}(\mathfrak{g}_\Sigma^*)/H_\Sigma$ consists of those points ξ in $\mathfrak{g}_\Sigma^*$ for which there is a z in P_Σ with $\mathbf{J}_{G_\Sigma}(z) = \mathbf{J}_G(z) = \xi$. By Corollary 8.1, this set is open in $\mathbf{J}_G^{-1}(\mathfrak{g}_\Sigma^*)/H$. ∎

Example Consider the situation of Example 1 of §8.3. We choose the same $P = \mathbb{C}^2$ and the same symplectic structure. However, now we choose $G = SU(2)$ and $H = U(1) = S^1$ with G acting by matrix multiplication and with H acting by the flow of two identical harmonic oscillators, with Hamiltonian given by $\mathcal{H}(z, w) = \frac{1}{2}(|z|^2 + |w|^2)$. These form a dual pair with the momentum maps given by the components of the Hopf map; see Cushman and Rod [1982] and Marsden [1987]. Again, we choose the discrete group Σ to be $\mathbb{Z}_2$ given by the map $\sigma(z, w) = (w, z)$. Here the hypotheses above are directly verified, and one finds, using the sort of computations in Example 1 of §8.3, that the discrete reduction of this dual pair is the trivial dual pair given by $P_\Sigma = \mathbb{C}^2$ with $G_\Sigma = S^1$ acting by complex multiplication and with $H_\Sigma = S^1$ acting the same way. In other words, the discrete reduction of the (completely integrable) system defined by the 1:1 resonance consisting of two identical harmonic oscillators (integrable via the group $SU(2)$), is the (completely integrable) system consisting of a single harmonic oscillator, trivially integrable using the group S^1. ◆

For more sophisticated examples of integrable systems, and in particular, the Kowalewski top of Bobenko, Reyman and Semenov-Tian-Shansky [1989] in this context, we refer the reader to Harnad, Hurtubise and Marsden [1991].

Chapter 9

Mechanical Integrators

For conservative mechanical systems with symmetry, it is of interest to develop numerical schemes that preserve this symmetry, so that the associated conserved quantities are preserved *exactly* by the integration process. One would also like the algorithm to preserve either the Hamiltonian or the symplectic structure — one cannot expect to do *both* in general, as we shall show below. There is some evidence (such as reported by Chanell and Scovel [1990] and Marsden et al. [1991]) that these mechanical integrators perform especially well for long time integrations, in which chaotic dynamics can be expected. It is well known that, in general, may standard algorithms can introduce spurious effects (such as nonexistent chaos) in long integration runs; see, for example, Reinhall, Caughey, and Storti [1989]. We use the general term *mechanical integrator* for an algorithm that respects one or more of the fundamental properties of being symplectic, preserving energy, or preserving the momentum map.

9.1 Definitions and Examples

By an *algorithm* on a phase space P we mean a collection of maps $F_\tau : P \to P$ (depending smoothly, say, on $\tau \in \mathbb{R}$ for small τ and $z \in P$). Sometimes we write $z^{k+1} = F_\tau(z^k)$ for the algorithm and we write Δt or h for the *step size* τ. We say that the algorithm is *consistent* or is *first order accurate* with a vector field X on P if

$$\left. \frac{d}{d\tau} F_\tau(z) \right|_{\tau=0} = X(z). \tag{9.1.1}$$

Higher order accuracy is defined similarly by matching higher order derivatives. One of the basic things one is interested in is *convergence* namely,

171

when is

$$\lim_{n\to\infty} (F_{t/n})^n(z) = \varphi_t(z) \tag{9.1.2}$$

where φ_t is the flow of X, and what are the error estimates? There are some general theorems guaranteeing this, with an important hypothesis being *stability*; *i.e.*, $(F_{t/n})^n(z)$ must remain close to z for small t and all $n = 1, 2, \ldots$. We refer to Chorin et al. [1978] and Abraham, Marsden and Ratiu [1988] for details. For example, the *Lie-Trotter formula*

$$e^{t(A+B)} = \lim_{n\to\infty} (e^{tA/n}e^{tB/n})^n \tag{9.1.3}$$

is an instance of this.

An algorithm F_τ is

1. a *symplectic-integrator* if each F_τ is symplectic,

2. an *energy-integrator* if $H \circ F_\tau = H$ (where $X = X_H$),

3. a *momentum-integrator* if $\mathbf{J} \circ F_\tau = \mathbf{J}$ (where $\mathbf{J}$ is the momentum map for a G-action).

If F_τ has one or more of these properties, we call it a *mechanical integrator*. Notice that if an integrator has one of these three properties, then so does any iterate of it.

There are two ways that have been employed to find mechanical integrators. For example, one can search amongst existing algorithms and find ones with special algebraic properties that make them symplectic or energy-preserving. Second, one can attempt to design mechanical integrators from scratch. Here are some simple examples:

Example 1 A first order explicit symplectic scheme in the plane is given by the map $(q_0, p_0) \mapsto (q, p)$ defined by

$$
\begin{aligned}
q &= q_0 + (\Delta t)p_0 \\
p &= p_0 - (\Delta t)V'(q_0 + (\Delta t)p_0). \tag{9.1.4}
\end{aligned}
$$

This map is a first order approximation to the flow of Hamilton's equations for the Hamiltonian $H = (p^2/2) + V(q)$. Here, one can verify by direct calculation that this scheme is in fact a symplectic map.   ♦

Example 1 is based on the use of generating functions, as we shall see below. A modification of Example 1 using Poincaré's generating function, but also one that can be checked directly is:

Example 2 An implicit symplectic scheme in the plane for the same Hamiltonian as in Example 1 is

$$
\begin{aligned}
q &= q_0 + (\Delta t)(p + p_0)/2 \\
p &= p_0 - (\Delta t)V'((q + q_0)/2). \quad \blacklozenge
\end{aligned}
\tag{9.1.5}
$$

Other examples are sometimes based on special observations. The next example shows that the second order accurate mid-point rule is symplectic (Feng [1987]). This algorithm is also useful in developing almost Poisson integrators (Austin, Krishnaprasad and Wang [1991]).

Example 3 In a symplectic vector space the following *mid point rule* is symplectic:

$$
\frac{z^{k+1} - z^k}{\Delta t} = X_H\left(\frac{z^k + z^{k+1}}{2}\right).
\tag{9.1.6}
$$

Notice that for small Δt the map defined implicitly by this equation is well defined by the implicit function theorem. To show it is symplectic, we use the fact that the *Cayley transform* S of an infinitesimally symplectic linear map A, namely

$$
S = (1 - \lambda A)^{-1}(1 + \lambda A)
\tag{9.1.7}
$$

is symplectic if $1 - \lambda A$ is invertible for some real λ. To apply this to our situation, rewrite the algorithm (9.1.6) as

$$
F_\tau(z) - z - \tau X_H\left(\frac{z + F_\tau(z)}{2}\right) = 0.
\tag{9.1.8}
$$

Letting $S = \mathbf{D}F_\tau(z)$ and $A = \mathbf{D}X_H\left(\frac{z + F_\tau(z)}{2}\right)$ we get, by differentiation, $S - 1 - \frac{1}{2}\tau A(1 + S) = 0$; *i.e.*, (9.1.7) holds with $\lambda = \tau/2$. Thus, (9.1.6) defines a symplectic scheme. $\quad \blacklozenge$

Example 4 Here is an example of an implicit energy preserving algorithm from Chorin et al. [1978].

Consider a Hamiltonian system for $\mathbf{q} \in \mathbb{R}^n$ and $\mathbf{p} \in \mathbb{R}^n$:

$$
\dot{\mathbf{q}} = \frac{\partial H}{\partial \mathbf{p}}, \quad \dot{\mathbf{p}} = -\frac{\partial H}{\partial \mathbf{q}}.
\tag{9.1.9}
$$

Define the following implicit scheme

$$
\mathbf{q}_{n+1} = \mathbf{q}_n + \Delta t \frac{H(\mathbf{q}_{n+1}, \mathbf{p}_{n+1}) - H(\mathbf{q}_{n+1}, \mathbf{p}_n)}{\lambda^T(\mathbf{p}_{n+1} - \mathbf{p}_n)}\lambda,
\tag{9.1.10}
$$

$$\mathbf{p}_{n+1} = \mathbf{p}_n - \Delta t \frac{H(\mathbf{q}_{n+1}, \mathbf{p}_n) - H(\mathbf{q}_n, \mathbf{p}_n)}{\mu^T(\mathbf{q}_{n+1} - \mathbf{q}_n)} \mu, \qquad (9.1.11)$$

where

$$\lambda = \frac{\partial H}{\partial \mathbf{p}}(\alpha \mathbf{q}_{n+1} + (1-\alpha)\mathbf{q}_n, \beta \mathbf{p}_{n+1} + (1-\beta)\mathbf{p}_n), \qquad (9.1.12)$$

$$\mu = \frac{\partial H}{\partial \mathbf{q}}(\gamma \mathbf{q}_{n+1} + (1-\gamma)\mathbf{q}_n, \delta \mathbf{p}_{n+1} + (1-\delta)\mathbf{p}_n), \qquad (9.1.13)$$

and where $\alpha, \beta, \gamma, \delta$ are arbitrarily chosen constants in $[0,1]$.

The proof of conservation of energy is simple: From (9.1.10), we have

$$(\mathbf{q}_{n+1} - \mathbf{q}_n)^T(\mathbf{p}_{n+1} - \mathbf{p}_n) = \Delta t(H(\mathbf{q}_{n+1}, \mathbf{p}_{n+1}) - H(\mathbf{q}_{n+1}, \mathbf{p}_n)), \quad (9.1.14)$$

and from (9.1.11)

$$(\mathbf{p}_{n+1} - \mathbf{p}_n)^T(\mathbf{q}_{n+1} - \mathbf{q}_n) = -\Delta t(H(\mathbf{q}_{n+1}, \mathbf{p}_n) - H(\mathbf{q}_n, \mathbf{p}_n)). \quad (9.1.15)$$

Subtracting (9.1.15) from (9.1.14), we obtain

$$H(\mathbf{q}_{n+1}, \mathbf{p}_{n+1}) = H(\mathbf{q}_n, \mathbf{p}_n). \qquad (9.1.16)$$

This algorithm is checked to be consistent. In general, it is not symplectic — this is in accord with Proposition 9.1 in the next section. ◆

Example 5 Let us apply the Lie-Trotter or time splitting idea to the simple pendulum. The equations are

$$\frac{d}{dt}\begin{pmatrix} \varphi \\ p \end{pmatrix} = \begin{pmatrix} p \\ 0 \end{pmatrix} + \begin{pmatrix} 0 \\ -\sin\varphi \end{pmatrix}.$$

Each vector field can be integrated explicitly to give maps

$$G_\tau(\varphi, p) = (\varphi + \tau p, p)$$

and

$$H_\tau(\varphi, p) = (\varphi, p - \tau \sin\varphi)$$

each of which is symplectic. Thus, the composition $F_\tau = G_\tau \circ H_\tau$, namely,

$$F_\tau(q, p) = (\varphi + \tau p - \tau^2 \sin\varphi, p - \tau \sin\varphi)$$

is a first order symplectic scheme for the simple pendulum. It is closely related to the *standard map*. The orbits of F_τ need not preserve energy and they may be chaotic, whereas the trajectories of the simple pendulum are of course not chaotic. ♦

We refer to the cited references, and to Ruth [1983], Feng [1986], Sanz-Serna [1988] and references therein for more examples of this type, including symplectic Runge-Kutta schemes.

9.2   Limitations on Mechanical Integrators

A number of algorithms have been developed specifically for integrating Hamiltonian systems to conserve the energy integral, but without attempting to capture all of the details of the Hamiltonian structure (see Example 4 above and also Stofer [1987] and Greenspan [1974, 1984]). In fact, some of the standard energy-conservative algorithms have poor momentum behavior over even moderate time ranges. This makes them unsuitable for problems in satellite dynamics for example, where the exact conservation of a momentum integral is essential to the control mechanism.

One can get angular momentum drift in energy-conservative simulations of, for example, rods that are free to vibrate and rotate, and presumably in the water molecule. To control such drifts and attain the high levels of computational accuracy demanded by automated control mechanisms, one would be forced to reduce computational step sizes to such an extent that the numerical simulation would be prohibitively inefficient. Similarly, if one attempts to use a standard energy-conservative algorithm to simulate both the rotational and vibrational modes of a freely moving flexible rod, the algorithm may predict that the rotational motion will come to a virtual halt after only a few cycles! For a documented simulation of a problem with momentum conservation, see Simo and Wong [1989]. One can readily imagine that in the process of enforcing energy conservation one could upset conservation of angular momentum.

What seems rather surprising is that all of the implicit members of the Newmark family, perhaps the most widely used time-stepping algorithms in nonlinear structural dynamics, *are not designed to conserve energy and also fail to conserve momentum.* Among the explicit members, only the central difference method preserves momentum. The analytitical proof of these results is in Simo, Tarnow and Wong [1991].

As we shall demonstrate in §9.4, the problem of how to generate a numerical algorithm that exactly conserves momentum (or, more generally, all momentum-like integrals of motion) is fairly easy to resolve. Since momentum integrals in Hamiltonian systems are associated with invariance of the system under the action of symmetry groups, one might guess that to derive momentum-conservative algorithms, one constrains the algorithm to obey, in some sense, the same group invariance as the actual dynamics.

In traditional integrators, much attention has been paid to energy conservation properties, some, as we have noted to momentum conservation, and even less to conserving the symplectic or Poisson structure. However, one can imagine that it is also quite important.

The three notions of symplectic, energy, and momentum integrators are connected in interesting ways. For example, as we shall show below, a result of Ge (see Ge and Marsden [1988]) is that under fairly weak additional assumptions, *a G-equivariant symplectic integrator is also momentum preserving.* For example, a symplectic integrator of this type applied to a free rigid body motion would exactly preserve the angular momentum vector in space.

Given the importance of conserving integrals of motion and the important role played by the Hamiltonian structure in the reduction procedure for a system with symmetry, one might hope to find an algorithm that combines all of the desirable properties: conservation of energy, conservation of momenta (and other independent integrals), and conservation of the symplectic structure. However, one cannot do all three of these things at once unless one relaxes one or more of the conditions in the following sense:

Proposition 9.1 *If one has an algorithm that is energy preserving, symplectic and momentum preserving and if the dynamics is nonintegrable on the reduced space (in the sense spelled out in the proof) then the algorithm already gives the exact solution of the original dynamics problem up to a time reparametrization.*

Proof Suppose $F_{\Delta t}$ is our symplectic algorithm of the type discussed above, and consider the application of the algorithm to the reduced phase space. We assume that the Hamiltonian H is the only integral of motion of the reduced dynamics (*i.e.,* all other integrals of the system have been found and taken out in the reduction process in the sense that any other conserved quantity (in a suitable class) is functionally dependent on H. Since $F_{\Delta t}$ is symplectic it is the Δt-time map of some *time-dependent* Hamiltonian function K. Now assume that the symplectic map $F_{\Delta t}$ also conserves H for *all* values of Δt. Thus $\{H, K\} = 0 = \{K, H\}$. The latter equation implies that K is functionally dependent on H since the flow of H (the "true dynamics") had no other integrals of motion. The functional dependence of K on H in turn implies that their Hamiltonian vector fields are parallel, so the flow of K (the approximate solution) and the flow of H (the exact solution) must lie along identical curves in the reduced phases space; thus the flows are equivalent up to time reparametrization. ■

Thus, it is unlikely one can find an algorithm that simultaneously conserves the symplectic structure, the momentum map, and the Hamiltonian. It is tempting (but probably wrong) to guess from this that one can monitor accuracy by keeping track of all three. Non-symplectic algorithms that conserve both momentum and energy have been studied by Simo and Wong [1989] and Austin, Krishnaprasad and Wang [1991]. We study the basic method in §9.5.

9.3 Symplectic Integrators and Generating Functions

Symplectic integrators based on generating functions have been developed by a large number of people, starting with de Vogelaere [1956] and Feng [1986]. We refer to Channell and Scovel [1990] for a survey.

Let us recall the following basic fact.

Proposition 9.2 *If* $S : Q \times Q \to \mathbb{R}$ *defines a diffeomorphism* $(q_0, p_0) \mapsto (q, p)$ *implicitly by*

$$p = \frac{\partial S}{\partial q} \quad and \quad p_o = -\frac{\partial S}{\partial q_0} \tag{9.3.1}$$

then this diffeomorphism is symplectic.

Proof Note that

$$\mathbf{d}S = \frac{\partial S}{\partial q^i} dq^i + \frac{\partial S}{\partial q_o{}^i} dq_o{}^i = p_i dq^i - p_{oi} dq_o{}^i$$

and so taking $\mathbf{d}$ of both sides, we get

$$dp_i \wedge dq^i = dp_{oi} \wedge dq_o{}^i$$

which means the diffeomorphism (9.3.1) is symplectic. ∎

Recall also one of the basic facts about Hamilton-Jacobi theory, namely that the flow of Hamilton's equations is the canonical transformation generated by the solution of the Hamilton-Jacobi equation

$$\frac{\partial S}{\partial t} + H\left(q, \frac{\partial S}{\partial q}\right) = 0 \tag{9.3.2}$$

where $S(q_0, q, t)|_{\substack{q=q_0 \\ t=0}}$ generates the identity. (This may require singular behavior in t. For example, $S = \frac{1}{2t}(q - q_0)^2$.) The strategy is to find an approximate solution of the Hamilton-Jacobi equation for small time Δt and to use this to obtain the algorithm using (9.3.1).

There are several other versions of the algorithm that one can also treat. For example, if specific coordinates are chosen on the phase space, one can use a generating function of the form $S(q^i, p_{0i}, t)$. In this case one can get the simple formula for a first order algorithm given in Example 1 in §9.1 by using $S = p_{0i}q^i - \Delta t H(q^i, p_{0i})$, which is easy to implement, and for Hamiltonians of the form kinetic plus potential, leads to the stated explicit symplectic algorithm.

9.4 Symmetric Symplectic Algorithms Conserve J

The construction of momentum-conserving algorithms, whether of symplectic or energy-momentum type, requires that level sets of the momentum map **J** remain invariant under the mapping $\varphi : P \to P$ that represents a single iteration of the algorithm. We next give sufficient conditions under which it is possible to obtain such a mapping in the symplectic case.

The argument is a modification of some ideas found in Ge and Marsden [1988]. We make the following assumptions:

   i P is a symplectic manifold endowed with an exact symplectic form $\omega = -\mathbf{d}\theta$;

   ii G is a Lie group acting symplectically on P and $\mathbf{J} : P \to \mathfrak{g}^*$ is an associated momentum map for the action, with g representing the action of $g \in G$;

   iii $\varphi : P \to P$ is a symplectic map;

   iv φ is G-equivariant: $\varphi(gz) = g\varphi(z)$, for all $z \in P$ and $g \in G$.

Letting $\xi_P = X_{\langle \mathbf{J}, \xi \rangle}$ designate the vector field corresponding to $\xi \in \mathfrak{g}$ under the action, we start by differentiating the equivariance condition

$$\varphi \circ g = g \circ \varphi$$

with respect to the group element in the direction of ξ at the identity of the group. This results in $\varphi^* \xi_P = \xi_P$ or, by definition of the momentum map,

$$\varphi^* X_{\langle \mathbf{J}, \xi \rangle} = X_{\langle \mathbf{J}, \xi \rangle}.$$

But $\varphi^* X_{\langle \mathbf{J}, \xi \rangle} = X_{\langle \mathbf{J}, \xi \rangle \circ \varphi}$ from assumption **iii**, so

$$X_{\langle \mathbf{J}, \xi \rangle \circ \varphi} = X_{\langle \mathbf{J}, \xi \rangle}.$$

Since two Hamiltonian vector fields are equal if and only if the Hamiltonians differ by a constant,

$$\langle \mathbf{J}, \xi \rangle \circ \varphi - \langle \mathbf{J}, \xi \rangle = \text{ constant}.$$

We need $\langle \mathbf{J}, \xi \rangle \circ \varphi = \langle \mathbf{J}, \xi \rangle$ for the value of $\mathbf{J}$ to be preserved by the map φ, so we need to establish sufficient conditions under which the constant will vanish.

We make the following further assumptions:

  v $S : P \rightarrow \mathbb{R}$ is a G-invariant generating function for the map φ; *i.e.*, $\varphi^* \theta = \theta + \mathbf{d}S$;

  vi $\langle \mathbf{J}, \xi \rangle = \mathbf{i}_{\xi_P} \theta$.

Then,

$$
\begin{aligned}
\langle \mathbf{J}, \xi \rangle \circ \varphi &= \varphi^* \langle \mathbf{J}, \xi \rangle = \varphi^* \mathbf{i}_{\xi_P} \theta \quad \text{(by \textbf{ii})} \\
&= \mathbf{i}_{\xi_P} \varphi^* \theta \quad \text{(by equivariance of } \varphi) \\
&= \mathbf{i}_{\xi_P} \theta + \mathbf{i}_{\xi_P} \mathbf{d}S \quad \text{(by \textbf{i})}.
\end{aligned}
$$

The first term in this last expression is $\langle \mathbf{J}, \xi \rangle$ again, and the final term vanishes by equivariance of S. Thus, the desired conservation condition, $\langle \mathbf{J}, \xi \rangle \circ \varphi = \langle \mathbf{J}, \xi \rangle$, follows from the assumptions.

Assuming that the original system is given in terms of canonical coordinates on a cotangent bundle $P = T^*Q$, we have $\omega = -\mathbf{d}\theta_0$, where $\theta_0 = p_i dq^i$ is the canonical one-form on the cotangent bundle. If the symmetry group G acts by cotangent lifts, then **vi** follows automatically. Condition **v** is equivalent to (9.1.1) if we regard S as a function of q_0 and q.

This argument applies to all "types" of generating functions, but when applied to ones of the form $S(q_0, q, t)$ we get:

Proposition 9.3 *Suppose that $S : Q \times Q \rightarrow \mathbb{R}$ is invariant under the diagonal action of G, i.e., $S(gq, gq_0) = S(q, q_0)$. Then the cotangent momentum map $\mathbf{J}$ is invariant under the canonical transformation φ_S generated by S, i.e., $\mathbf{J} \circ \varphi_S = \mathbf{J}$.*

This may also be seen directly by differentiating the invariance condition assumed on S with respect to $g \in G$ in the direction of $\xi \in \mathfrak{g}$ and utilizing the definitions of $\varphi_S, \mathbf{J}$ and ξ_Q. The following is also true: *If G acts on Q freely, and a given canonical transformation φ conserves $\mathbf{J}$, then its generating function S can be defined on an open set of $Q \times Q$ which is invariant under the action of G, and S is invariant under the action of G.* This is proved in Ge [1991a].

Note that if H is invariant under the action of G, then the corresponding solution of the Hamilton-Jacobi equation is G invariant as well. This follows from the short time uniqueness of the generating function of the type assumed for the flow of the Hamiltonian vector field X_H determined by H. It also follows from Proposition 9.1 that if the approximate solution of the Hamilton-Jacobi equation is chosen to be G-invariant, then the corresponding algorithm will exactly conserve the momentum map.

For a start on the numerical analysis of symplectic integrators, see Sanz-Serna [1988], Simo, Tarnow, and Wong [1991] and related papers. This whole area needs further development for the community to be able to intelligently choose amongst various algorithms. For instance, from the point of view of stability, the optimal second-order accurate symplectic integrators are the mid point rule and the central difference method.

9.5   Energy-Momentum Algorithms

We now turn to some basic remarks on the construction of algorithms that conserves the Hamiltonian and the momentum map, but will not, in general, conserve the symplectic structure.

A class of algorithms satisfying this requirement can be obtained through the steps outlined below. The geometry of the process is depicted in Figure 9.5.1.

i   Formulate an energy-preserving algorithm on the symplectic reduced phase space $P_\mu = \mathbf{J}^{-1}(\mu)/G_\mu$ or the Poisson reduced space P/G. If such an algorithm is interpreted in terms of the primitive phase space P, it becomes an iterative mapping from one orbit of the group action to another.

ii   In terms of canonical coordinates (q, p) on P, interpret the orbit-to-orbit mapping described above and if P/G was used, impose the constraint $\mathbf{J}(q_k, p_k) = \mathbf{J}(q_{k+1}, p_{k+1})$. The constraint does not uniquely determine the restricted mapping, so we may obtain a large class if iterative schemes.

iii To uniquely determine a map from within the above class, we must determine how points in one G_μ-orbit are mapped to points in another orbit. There is still an ambiguity about how phase space points drift in the G_μ-orbit directions. This drift is closely connected with *geometric phases* (Chapter 6)! In fact by discretizing the geometric phase formula for the system under consideration we can specify the shift along each G_μ-orbit associated with each iteration of the map.

The papers of Simo and Wong [1989] and Krishnaprasad and Austin [1990] provide systematic methods for making the choices required in steps

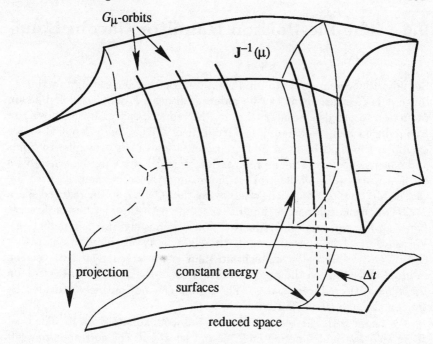

Figure 9.5.1: An energy preserving algorithm is designed on the reduced space and is then lifted to the level set of the momentum map by specifying phase information.

ii and **iii**. The general construction given above is, in fact, precisely the approach advocated in ref Simo, Tarnow and Wong [1991]. There it is shown that projection from the level set of constant angular momentum onto the surface of constant energy can be performed implicitly or explicitly leading to predictor/corrector type of algorithms. From a numerical analysis standpoint, the nice thing is that the cost involved in the actual construction of the projection reduces to that of a line search (i.e., basically for free). The algorithm advocated in Simo and Wong [1989] is special in the sense that the projection is not needed for $Q = SO(3)$: the discrete flow is shown to lie in the intersection of the level set of angular momentum and the surface of constant energy. This algorithm is singularity–free and integrates the dynamics exactly up to a time reparametrization, consitent with the restrictions on mechanical integrators given in §9.2. Extensions of these schemes to elasticity, rods and shells suitable for large-scale calculation and amenable to paralelization are given in Simo, Fox and Rifai [1991], and Simo and Doblare [1991].

9.6 The Lie-Poisson Hamilton-Jacobi Equation

Suppose that we are able to produce a G-invariant generating function S. Since it is G-invariant, it can be reduced, either by symplectic or Poisson reduction to produce an algorithm on the reduced space. It also gives rise to a reduced version of Hamilton-Jacobi theory. This can be applied to, for example, the rigid body in body representation or, in principle, to fluids and plasmas in the spatial representation. Instead of giving the generalities of the theory, we shall illustrate it in an important case, namely, with the case of Lie-Poisson reduction, whereby we take $Q = G$, so the reduced space T^*Q/G is isomorphic with the dual space $\mathfrak{g}^*$ with the Lie-Poisson bracket (with a plus sign for right reduction and a minus sign for left reduction). We shall give the special case of the rigid body for illustration, taking $G = SO(3)$. Since the momentum map is preserved, one also gets an induced algorithm on the coadjoint orbits, or in the more general cases, on the symplectic reduced spaces. The proofs can be routinely provided by tracing through the definitions.

We begin with the reduced Hamilton-Jacobi equation itself. Thus, let H be a G-invariant function on T^*G and let H_L be the corresponding left reduced Hamiltonian on $\mathfrak{g}^*$. (To be specific, we deal with left actions — of course there are similar statements for right reduced Hamiltonians.) If S is invariant, there is a unique function S_L such that $S(g, g_0) = S_L(g^{-1}g_0)$. (One gets a slightly different representation for S by writing $g_0^{-1}g$ in place of $g^{-1}g_0$.)

Proposition 9.4 *The left reduced Hamilton-Jacobi equation is the following equation for a function* $S_L : G \to \mathbb{R}$:

$$\frac{\partial S_L}{\partial t} + H_L(-TR_g^* \cdot \mathbf{d}S_L(g)) = 0 \qquad (9.6.1)$$

*which we call the **Lie-Poisson Hamilton-Jacobi equation**. The Lie-Poisson flow of the Hamiltonian* H_L *is generated by the solution* S_L *of (9.6.1) in the sense that the flow is given by the Poisson transformation of* $\mathfrak{g}^*$: $\Pi_0 \mapsto \Pi$ *defined as follows: Define* $g \in G$ *by solving the equation*

$$\Pi_0 = -TL_g^* \cdot \mathbf{d}_g S_L \qquad (9.6.2)$$

for $g \in G$ *and then setting*

$$\Pi = Ad_{g^{-1}}^* \Pi_0. \qquad (9.6.3)$$

Here *Ad* denotes the adjoint action and so the action in (9.6.3) is the coadjoint action. Note that (9.6.3) and (9.6.2) give $\Pi = -TR_g^* \cdot dS_L(g)$. Note also that (9.6.2) and (9.6.3) are the analogues of Equation (9.1.1) and that (9.6.1) is the analogue of (9.1.2). Thus, one can obtain a Lie-Poisson integrator by approximately solving (9.6.1) and then using (9.6.2) and (9.6.3) to generate the algorithm. This algorithm (9.6.3) manifestly preserves the coadjoint orbits (the symplectic leaves in this case). As in the canonical case, one can generate algorithms of arbitrary accuracy this way.

There may be conditions necessary on Π_0 for the solvability of Equation (9.6.2). This is noted in the example of the rigid body below.

For the case of the rigid body, these equations read as follows. First, Equation (9.6.1) reads

$$\frac{\partial S_L}{\partial t} + H_L(-\nabla S_L(A) \cdot A^T) = 0 \qquad (9.6.4)$$

i.e.,

$$\frac{\partial S_L}{\partial t} + H_L\left(-\frac{\partial S_L}{\partial A^i{}_j} A^k{}_j\right) \qquad (9.6.5)$$

(sum over j) where the action function S_L is a function of an orthogonal matrix A and where we have identified tangent and cotangent spaces using the bi-invariant metric on the rotation group. This metric corresponds to the standard Euclidean metric on the Lie algebra, when identified with Euclidean 3-space. This identification maps the Lie algebra bracket to the cross product. The expression $\nabla S_L(A) \cdot A^T$ is a skew symmetric matrix, *i.e.,* it lies in the Lie algebra $so(3)$, so it makes sense for H_L to be evaluated on it. As usual, one has to be careful how the gradient (derivative) ∇S_L is computed, since there is a constraint $AA^T = I$ involved. If it is computed naively in the coordinates of the ambient space of 3×3 matrices, then one interprets the expression $\nabla S_L(A) \cdot A^T$ using naive partial derivatives and skew symmetrizing the result; this projects the gradient to the constraint space, so produces the gradient of the constrained function.

Equation (9.6.5) thus is the Hamilton-Jacobi equation for the dynamics of a rigid body *written directly in body representation*. The flow of the Hamiltonian is generated by S_L in the following way: It is the transformation of initial conditions at time $t = 0$ to a general t determined by first solving the equation

$$\hat{\Pi}_0 = -A^T \cdot \nabla S_L(A) \qquad (9.6.6)$$

for the matrix A and then setting $\Pi = A\Pi_0$, where $\hat{\Pi} = [\Pi^i{}_j]$ is the skew *matrix* associated to the *vector* Π in the usual way: $\hat{\Pi} \cdot v = \Pi \times v$. (Again, the right hand side of (9.6.6) is to be skew symmetrized if the derivative was taken in the naive way with the constraint ignored.) We have written the result in terms of the body angular momentum vector Π; one can rewrite it in terms of the body angular velocity vector by using the relation $\Pi = I\omega$, where I is the moment of inertia tensor. In coordinates, Equation (9.6.6) reads as follows:

$$(\Pi_0)^k_i = -A^j{}_i \frac{\partial S_L}{\partial A^j{}_k}. \tag{9.6.7}$$

Finally, we note that similar equations also apply for fluids and plasmas, since they are also Lie-Poisson systems (but with right reduction). Also, the methods here clearly will generalize to the situation for reduction of any cotangent bundle; this generality is needed for example, for the case of free boundary fluids — see Lewis, Marsden, Montgomery and Ratiu [1986]. One of the ideas of current interest is to use mechanical integrators on vortex algorithms. For example, one can use it on point vortices in the plane (Pullen and Saffman [1991]) or on vortex dipoles in three space (cf. Rouhi [1988] and Buttke [1991]) both of which live on finite dimensional coadjoint orbits for the Euler equations (Marsden and Weinstein [1983]).

A general way to construct first order algorithms valid in the Lie-Poisson setting (as well as its analogues in the symplectic and Poisson context) is as follows. Let $H : \mathfrak{g}^* \to \mathbb{R}$ be a given Hamiltonian function and let S_0 be a function that generates a Poisson transformation $\varphi_0 : \mathfrak{g}^* \to \mathfrak{g}^*$ and let

$$S_{\Delta t} = S_0 + \Delta t H(L^*_g dS_0). \tag{9.6.8}$$

For small Δt, (9.6.8) generates a Poisson transformation, say $\varphi_{\Delta t} : \mathfrak{g}^* \to \mathfrak{g}^*$. Then we have:

Proposition 9.5 *With the assumptions above, the algorithm*

$$\Pi^k \mapsto \Pi^{k+1} = \varphi_0^{-1} \circ \varphi_{\Delta t}(\Pi^k) \tag{9.6.9}$$

is a Poisson difference scheme that is a first order difference scheme for the Hamiltonian system with Hamiltonian H.

In particular, if one can generate the identity transformation with a function S_0, then one can get a specific first order scheme. On G, one can introduce singularities in the time variable to do this, as we have already remarked. Interestingly, for $\mathfrak{g}$ semisimple, one can do this in a ***non-singular*** way on $\mathfrak{g}^*$. In fact, in this case, the function

$$S_0(g) = \text{trace}(Ad^*_g) \tag{9.6.10}$$

generates the identity in a G-invariant neighborhood of the zero of $\mathfrak{g}^*$. One can also check this with a direct calculation using (9.6.2) and (9.6.3). The neighborhood condition is necessary since there may be some restrictions on Π_0 required for the solvability of (9.6.2). For example, for the rigid body the condition is checked to be $\|\Pi_0\| < 1$. This condition can be dealt with using a scaling argument. We note that when one solves (9.6.2) for g, it need not be the identity, and consistent with (9.6.3) we observe that the solution g lies in the coadjoint isotropy of the element Π_0.

More generally, we say that a Lie algebra is a *regular quadratic Lie algebra* if there is a symmetric, Ad-invariant, non-degenerate bilinear form on $\mathfrak{g}$.

Proposition 9.6 *There is a function S_0 defined in some neighborhood of the identity element in G that generates the identity map of $\mathfrak{g}^*$ iff $\mathfrak{g}$ is a regular quadratic Lie algebra.*

Proof If S_0 exists, define the bilinear form $B : \mathfrak{g} \times \mathfrak{g} \to \mathbb{R}$ by

$$B(\xi, \eta) = \left.\frac{d}{ds}\right|_{s=0} \left.\frac{d}{dt}\right|_{t=0} S_0(\exp(s\xi)\exp(t\eta))$$

for $\xi, \eta \in \mathfrak{g}$. One verifies B is Ad-invariant and is non-degenerate from the fact that S_0 generates the identity. Conversely, given B, define S_0 by

$$S_0(g) = B(\log g, \log g) \tag{9.6.11}$$

where $\log$ is a local inverse of the exponential map. ∎

Combining (9.6.8) and (9.6.10) we get the following proposition.

Proposition 9.7 *The generating function*

$$S_{\Delta t}(g) = \text{trace}(Ad_g^*) + \Delta t H(L_g^* \mathbf{d}\, \text{trace}(Ad_g^*)). \tag{9.6.12}$$

defines, via (9.6.2) and (9.6.3), a Poisson map that is a first order Poisson integrator for the Hamiltonian H.

We remark that this scheme will automatically preserve additional conserved quantities on $\mathfrak{g}^*$ that, for example, arise from invariance of the Hamiltonian under a subgroup of G acting on the *right*. This is the situation for a rigid body with symmetry and fluid flow in a symmetric container (with left and right swapped), for instance.

9.7 Example: The Free Rigid Body

For the case of the free rigid body, we let $so(3)$, the Lie algebra of $SO(3)$, be the space of skew symmetric 3×3 matrices. An isomorphism between $so(3)$ and $\mathbb{R}$ is given by mapping the skew vector v to the matrix $\hat{v}$ defined previously. Using the Killing form $\langle A, B \rangle = \frac{1}{2} \operatorname{trace} A^T B$, which corresponds to the standard inner product on $\mathbb{R}^3$, *i.e.*, $\langle \hat{v}, \hat{w} \rangle = v \cdot w$, we identify $so(3)$ with $so(3)^*$. We write the Hamiltonian $H : so(3) \to \mathbb{R}$ as $H(\hat{v}) = \frac{1}{2} v \cdot Iv$, where I is the moment of inertia tensor. Let $\hat{I} : so(3) \to so(3)$ be defined by $\hat{I}(\hat{v}) = (Iv)\hat{\ }$. Thus, $H(\hat{v}) = \frac{1}{4} \langle \hat{v}, \hat{I}(\hat{v}) \rangle$. Equation (9.6.10) becomes $S_0(A) = \operatorname{trace}(A)$ and so $TL^*_A dS_0 = \frac{1}{2}(A - A^T)$. Therefore, (9.6.12) becomes

$$S_{\Delta t}(A) = \operatorname{trace}(A) + \Delta t H \left(\frac{1}{2}(A - A^T) \right) \tag{9.7.1}$$

and so Proposition 9.5 gives the following specific Lie-Poisson algorithm for rigid body dynamics: It is the scheme $\Pi^k \mapsto \Pi^{k+1}$ defined by

$$\hat{\Pi}^k = \frac{1}{2} \left[\frac{1}{4} \{ A\hat{I}(A - A^T) + \hat{I}(A - A^T)A^T \} \Delta t + (A - A^T) \right] \tag{9.7.2}$$

$$\hat{\Pi}^{k+1} = \frac{1}{2} \left[\frac{1}{4} \{ \hat{I}(A - A^T)A + A^T \hat{I}(A - A^T) \} \Delta t + (A - A^T) \right] \tag{9.7.3}$$

where, as before, the first equation is to be solved for the rotation matrix A and the result substituted into the second. Letting $A^S = \frac{1}{2}[AA^T]$ denote the skew part of the matrix A, we can rewrite the scheme as

$$\hat{\Pi}^k = A^S + (A\hat{I}A^S)^S \Delta t \tag{9.7.4}$$

$$\hat{\Pi}^{k+1} = A^S + (A^T \hat{I}A^S)^S \Delta t. \tag{9.7.5}$$

Of course, one can write $A = \exp(\xi)$ and solve for ξ and express the whole algorithm in terms of $\mathfrak{g}$ and $\mathfrak{g}^*$ alone. This type of algorithm does not keep track (except implicitly) of the rigid body phase. One can imagine combining the ideas here with those in Chapter 6 to do that.

We know from the general theory that this scheme will automatically be Poisson and will, in particular, preserve the coadjoint orbits, *i.e.*, the total angular momentum surfaces $\|\Pi\|^2 = $ constant. Of course, using other choices of S_0, it is possible to generate other algorithms for the rigid body, but the choice $S_0(A) = \operatorname{trace}(A)$ is particularly simple. We point out the interesting feature that the function (9.6.10) for the case of ideal Euler fluid flow is the function that assigns to a fluid placement field φ (an element of the diffeomorphism group of the containing region) the trace of the linear operator $\omega \mapsto \varphi^*\omega$, on vorticity fields ω, which measures the vortex distortion due to the nonrigidity of the flow. (See Marsden and Weinstein [1983] for further information.)

9.8 Variational Considerations

We mention briefly another approach to obtaining symplectic integrators that shows promise of yielding information on the behavior of energy (why it can be conserved *on average* in symplectic integrators). The approach here is based on work with S. Weissman.

As we saw in Chapter 2, the usual variational principle, which is equivalent to the Euler-Lagrange equations, is

$$\delta \int_a^b L\, dt = 0 \tag{9.8.1}$$

where the endpoints are *fixed*. Now modify (9.8.1) to

$$\delta \int \left[L - \frac{d}{dt} S \right] dt = 0 \tag{9.8.2}$$

where the endpoints are *free*; here S is a function of q, q_0, and t. Condition (9.8.2) gives (suppressing indices and assuming S is constant when $q = q_0$, so it does not contribute to the lower limit),

$$\int \left(\frac{\partial L}{\partial q} \delta q + \frac{\partial L}{\partial \dot{q}} \delta \dot{q} \right) dt - \delta(S(q, q_0, t)) = 0$$

i.e.,

$$\int \left(\frac{\partial L}{\partial q} - \frac{d}{dt} \frac{\partial L}{\partial \dot{q}} \right) \delta q\, dt + p\delta q - p_0 \delta q_0 - \frac{\partial S}{\partial q} \delta q - \frac{\partial S}{\partial q_0} \delta q_0 = 0. \tag{9.8.3}$$

Proposition 9.8 *The variational principle (9.8.2) with free endpoints is equivalent to the Euler-Lagrange equations plus the conditions (9.1.1) defining a symplectic map.*

This principle works the same way starting using the phase space variational principle

$$\delta \int \left\{ p_i \dot{q}^i - H(q, p) - \frac{d}{dt} S(q, q_0, t) \right\} dt. \tag{9.8.4}$$

Thus we see that the symplectic nature of the flow can be built into the variational principle. We now imagine S is determined by requiring it be an (approximate) solution of the Hamilton-Jacobi equation, which will make the symplectic map (9.1.1) compatible with the Euler-Lagrange equations.

If we discretize the principle (9.8.2) or (9.8.4) using finite element ideas in time, then we can take a basis of functions to describe $q(t)$ (or $(q(t), p(t))$ for (9.8.4) and enforce the principle on this set, ensuring enough freedom so the boundary conditions are met exactly. Also, conservation of energy (in a weak or averaged sense) can be viewed as one of the conditions that determines S. As before, we also want S to be G-invariant to ensure conservation of the momentum map.

An interesting way to introduce the Hamilton-Jacobi equation is to observe that *the equations of motion are consistent with* (9.3.1) *iff S satisfies the Hamilton-Jacobi equation* (with H possibly modified by a constant).

Chapter 10

Hamiltonian Bifurcation

In this chapter, we study some examples of bifurcations in the Hamiltonian context. A lot of the ideas from the previous chapters come into this discussion, and links with new ones get established, such as connections with chaotic dynamics and solution spaces in relativistic field theories. Our discussion will be by no means complete; it will focus on certain results of personal interest and results that fit in with the rest of the chapters. Some additional information on bifurcation theory in the Hamiltonian context may be found in the references cited below and in Abraham and Marsden [1978], Arnold [1978], Meyer and Hall [1991] and the references therein.

10.1 Some Introductory Examples

Bifurcation theory deals with the changes in the phase portrait structure of a given dynamical system as parameters are varied. One usually begins by focussing on the simplest features of the phase portrait, such as equilibrium points, relative equilibria, periodic orbits, relative periodic orbits, homoclinic orbits, etc., and studies how they change in number and stability characteristics as the system parameters are changed. Often these changes lead to new structures, such as more equilibria, periodic orbits, tori, or chaotic solutions, and the way in which stability or instability is transfered to these new structures from the old ones is of interest.

As we pointed out in Chapter 2, the symmetry (isotropy) group of a point in phase space determines how degenerate it is for the momentum map. Correspondingly, one expects (see Golubitsky, Sheaffer and Stewart [1988]), that these symmetry groups will play a vital role in the bifurcation theory of relative equilibria and its connections with dynamic stability

theory. The beginnings of this theory have started and as it evolves, it
will be tightly tied with the normal form methods and with the topology
of the level sets of $H \times \mathbf{J}$ and their associated bifurcations as the level set
values and other system parameters vary. To begin, one can look at the
case in which the symmetry group of a point is discrete, so that the point
is a regular value of the momentum map. Already this case is reasonably
rich; we shall comment on the more general case in §10.5 below. In the case
of discrete symmetry groups, the energy momentum method can be used
to help put the linearized system into normal form and from this one can
calculate the stability transitions, and the eigenvalue evolution.

To begin with a simple example, consider a ball moving with no friction
in a hoop constrained to rotate with angular velocity ω (Figure 10.1.1).

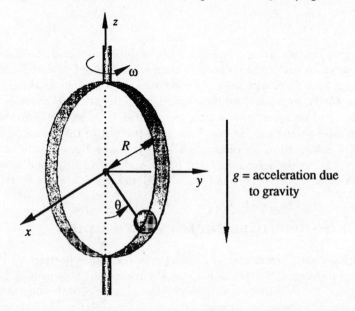

Figure 10.1.1: A ball in a rotating hoop.

In a moment we shall show that as ω increases past $\sqrt{g/R}$, the stable
equilibrium at $\theta = 0$ becomes unstable through a *Hamiltonian pitchfork
bifurcation* (Figure 10.1.2). The symmetry of these phase portraits is a
reflection of the original $\mathbb{Z}_2$ symmetry of the mechanical system. One can
break this symmetry by, for example, putting the rotation axis slightly off-
center, as we shall discuss below. Breaking this symmetry is an example
of *system symmetry breaking* since it is the whole system that loses
the symmetry. In Figure 10.1.2 notice that the stable *solution* on the left

has $\mathbb{Z}_2$ symmetry, while those on the right do not. This is an example of *solution symmetry breaking* within a symmetric system.

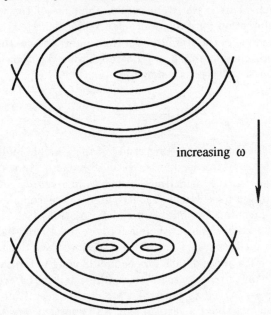

increasing ω

Figure 10.1.2: Hamiltonian pitchfork bifurcation as ω passes criticality.

This example will be compared with bifurcations of a planar liquid drop (with a free boundary held with a surface tension τ) following Lewis, Marsden and Ratiu [1987] and Lewis [1989]. In this example, a circular drop loses its circular symmetry to a drop with $\mathbb{Z}_2 \times \mathbb{Z}_2$ symmetry as the angular momentum of the drop is increased (although the stability analysis near the bifurcation is somewhat delicate). There are also interesting stability and bifurcation results in the dynamics of vortex patches, especially those of Wan in a series of papers starting with Wan and Pulvirente [1984].

The ball in the hoop is an example of a *steady state bifurcation*. This situation is already complicated, even when only discrete symmetry of the underlying system is considered. A similar bifurcation occurs in the dynamics of rotating planar coupled rigid bodies, as was analyzed by Oh, Sreenath, Krishnaprasad and Marsden [1989] and in the bifurcations of two coupled rigid bodies, as in Patrick [1991]. It also occurs in our double spherical pendulum, as we shall see.

Another basic bifurcation in the Hamiltonian context is the *1:1 resonance*, or *the Hamiltonian Hopf bifurcation*. In this case, two eigenvalues of the system linearized at a given equilibrium (or relative equilibrium) come together on the imaginary axis — that is, two frequencies in the system become equal. This is the Hamiltonian analogue of the well known Poincaré-Hopf bifurcation for nonhamiltonian vector fields.

A classical example of the Hamiltonian Hopf bifurcation is the fast slow transition in an upright spinning heavy Lagrange top. When the top makes the transition from a fast top to a slow top

$$\omega \downarrow \frac{2\sqrt{Mg\ell I_1}}{I_3}$$

an instability sets in of this sort. For a study of the possible bifurcations in the dynamics of a heavy top see Lewis, Ratiu, Simo and Marsden [1992].

Behavior of this sort is sometimes called a *gyroscopic instability*, or a *Krein collision* (see Krein [1950]). Here more complex dynamic behavior ensues, including periodic and chaotic motions (see also Holmes and Marsden [1983] and references therein for how chaotic motion is related to a homoclinic orbit that grows out of the fast-slow transition).

Next, we give a few more details for the ball moving in a rotating hoop. The particle is assumed to have mass m and be acted on by gravitational and frictional forces, as well as constraint forces that keep it on the hoop. The hoop itself is spun about a vertical axis with constant angular velocity ω, as in Figure 10.1.3.

The position of the ball *in space* is specified by the angles θ and φ, as shown. We can take $\varphi = \omega t$, so the position of the ball becomes determined by θ alone. Let the orthonormal frame along the coordinate directions e_θ, e_φ and e_r be as shown. The forces acting on the particle are:

1. Friction, proportional to the velocity of the ball relative to the hoop: $-\nu R \dot\theta e_\theta$, where $\nu \geq 0$ is a constant.

2. Gravity: $-mg\mathbf{k}$.

3. Constraint forces in the directions e_r and e_φ to keep the ball in the hoop.

The equations of motion are derived from Newton's second law $\mathbf{F} = m\mathbf{a}$. To get them, we calculate the acceleration $\mathbf{a}$. Relative to the xyz coordinate system, we have

$$\left. \begin{array}{rcl} x &=& R\sin\theta\cos\varphi \\ y &=& R\sin\theta\sin\varphi \\ z &=& -R\cos\theta \end{array} \right\} . \tag{10.1.1}$$

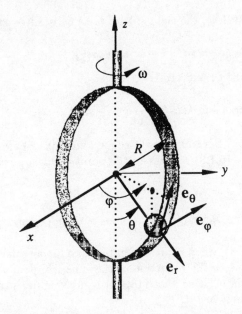

Figure 10.1.3: Coordinates for the derivation of the equations.

Calculating the second derivatives using $\varphi = \omega t$ and the chain rule in (10.1.1) gives

$$
\begin{aligned}
\ddot{x} &= -\omega^2 x - \dot{\theta}^2 x + (R\cos\theta\cos\varphi)\ddot{\theta} - 2R\omega\dot{\theta}\cos\theta\sin\varphi \\
\ddot{y} &= -\omega^2 y - \dot{\theta}^2 y + (R\cos\theta\sin\varphi)\ddot{\theta} + 2R\omega\dot{\theta}\cos\theta\cos\varphi \quad (10.1.2) \\
\ddot{z} &= -z\dot{\theta}^2 + R\sin\theta\,\ddot{\theta}.
\end{aligned}
$$

If $\mathbf{i}, \mathbf{j}, \mathbf{k}$ denote unit vectors along the x, y, and z axes respectively, then

$$
\mathbf{e}_\theta = (\cos\theta\cos\varphi)\mathbf{i} + (\cos\theta\sin\varphi)\mathbf{j} + \sin\theta\mathbf{k}. \qquad (10.1.3)
$$

In $\mathbf{F} = m\mathbf{a}$, $\mathbf{F}$ is the sum of the three forces described earlier and

$$
\mathbf{a} = \ddot{x}\mathbf{i} + \ddot{y}\mathbf{j} + \ddot{z}\mathbf{k}. \qquad (10.1.4)
$$

The $\mathbf{e}_\varphi$ and $\mathbf{e}_r$ components of $\mathbf{F} = m\mathbf{a}$ tell us what the constraint forces must be; the equation of motion comes from the $\mathbf{e}_\theta$ component:

$$
\mathbf{F} \cdot \mathbf{e}_\theta = m\mathbf{a} \cdot \mathbf{e}_\theta. \qquad (10.1.5)
$$

Using (10.1.3), the left side of (10.1.5) is

$$
\mathbf{F} \cdot \mathbf{e}_\theta = -\nu R\dot{\theta} - mg\sin\theta \qquad (10.1.6)
$$

while from (10.1.1) – (10.1.4), the right side of (10.1.5) is, after some algebra,

$$\mathbf{ma} \cdot \mathbf{e}_\theta = mR\{\ddot{\theta} - \omega^2 \sin\theta cos\theta\}. \tag{10.1.7}$$

Comparing (10.1.5), (10.1.6) and (10.1.7), we get

$$\ddot{\theta} = \omega^2 \sin\theta\cos\theta - \frac{\nu}{m}\dot{\theta} - \frac{g}{R}\sin\theta \tag{10.1.8}$$

as the equation of motion. Several remarks concerning (10.1.8) are in order:

i   If $\omega = 0$ and $\nu = 0$, (10.1.8) reduces to the ***pendulum equation***

$$R\ddot{\theta} + g\sin\theta = 0. \tag{10.1.9}$$

In fact our system can be viewed just as well as a ***whirling planar pendulum***.

We notice that (10.1.8), when expressed in terms of the angular momentum is *formally* the same as the reduction of the spherical pendulum. See Equation (3.5.9). It is interesting that (10.1.8) is nonsingular near $\theta = 0$, whereas (3.5.9) has a singularity.

ii   For $\nu = 0$, (10.1.8) is Hamiltonian with respect to $q = \theta, p = mR^2\dot{\theta}$, the canonical bracket structure

$$\{F, K\} = \frac{\partial F}{\partial q}\frac{\partial K}{\partial p} - \frac{\partial K}{\partial q}\frac{\partial F}{\partial p}$$

and energy

$$H = \frac{p^2}{2m} - mgR\cos\theta - \frac{mR^2\omega^2}{2}\sin^2\theta. \tag{10.1.10}$$

We can also use Lagrangian methods to derive (10.1.8). From the figure, the velocity is $v = R\dot{\theta}\mathbf{e}_\theta + (\omega R\sin\theta)\mathbf{e}_\varphi$, so

$$T = \frac{1}{2}m\|\mathbf{v}\|^2 = \frac{1}{2}m(R^2\dot{\theta}^2 + [\omega R\sin\theta]^2) \tag{10.1.11}$$

while the potential energy is,

$$V = -mgR\cos\theta, \tag{10.1.12}$$

so we choose

$$L = T - V = \frac{1}{2}mR^2\dot{\theta}^2 + \frac{mR^2\omega^2}{2}\sin^2\theta + mgR\cos\theta. \tag{10.1.13}$$

The Euler-Lagrange equations

$$\frac{d}{dt}\frac{\partial L}{\partial \dot{\theta}} - \frac{\partial L}{\partial \theta} = T$$

then give (10.1.8). The Legendre transform gives $p = mR^2\dot{\theta}$ and the Hamiltonian (10.1.10).

Consider *equilibrium solutions*; *i.e.*, solutions satisfying $\dot{\theta} = 0$, and $\ddot{\theta} = 0$; (10.1.8) gives

$$R\omega^2 \sin\theta \cos\theta = g\sin\theta. \tag{10.1.14}$$

Certainly $\theta = 0$, and $\theta = \pi$ solve (10.1.14) corresponding to the particle at the bottom or top of the hoop. If $\theta \neq 0$ or π, (10.1.14) becomes

$$R\omega^2 \cos\theta = g \tag{10.1.15}$$

which has two solutions when $g/(R\omega^2) < 1$. The value

$$\omega_c = \sqrt{\frac{g}{R}} \tag{10.1.16}$$

is the *critical rotation rate*. (Notice that ω_c is the frequency of linearized oscillations for the simple pendulum *i.e.*, for $R\ddot{\theta}+g\theta = 0$.) For $\omega < \omega_c$ there are only *two* solutions $\theta = 0, \pi$, while for $\omega > \omega_c$ there are *four solutions*,

$$\theta = 0, \pi, \pm\cos^{-1}\left(\frac{g}{R\omega^2}\right). \tag{10.1.17}$$

We say that a *Hamiltonian pitchfork bifurcation* occurs as ω crosses ω_c.

This system with $\nu = 0$ is symmetric in the sense that the symplectic $\mathbb{Z}_2$-action given by $\theta \mapsto -\theta$, and $\dot{\theta} \mapsto -\dot{\theta}$ leaves the phase portrait invariant. If this $\mathbb{Z}_2$ symmetry is broken, by setting the rotation axis a little off center, for example, then one side gets preferred, as in Figure 10.1.4. Let ϵ denote the off-center distance, so ϵ is a *symmetry breaking parameter*.

The evolution of the phase portrait for $\epsilon \neq 0$ is shown in Figure 10.1.5.

Near $\theta = 0$, the potential function has changed from the symmetric bifurcation in Figure 10.1.2 to the unsymmetric one in Figure 10.1.6. This is the *cusp catastrophe*.

Aside For the symmetric ball in the hoop, imagine that the hoop is subject to small periodic pulses; say $\omega = \omega_0 + \epsilon\cos(\eta t)$, or perhaps the hoop is coupled to another oscillator. Using the Melnikov method described below, it is reasonable to expect that the resulting time periodic system has horseshoe chaos if ϵ and ν are both small, but ϵ/ν is large enough.

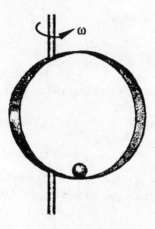

Figure 10.1.4: The ball in the off-center hoop.

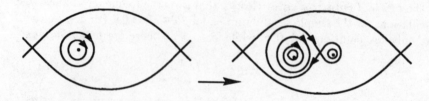

Figure 10.1.5: Phase portrait for the ball in the off-center hoop.

10.2   The Role of Symmetry

Consider a Hamiltonian vector field X_H on a phase space P depending on a parameter λ (like the angular velocity of the hoop in Figure 10.1.1) and we have a given curve $z(\lambda)$ of equilibrium solutions — these can also be *relative* equilibria if we work on the reduced phase space. If we linearize the vector field at the equilibrium we get a linear Hamiltonian system on the tangent space $T_{z(\lambda)}P$ and we can examine its eigenvalues. For relative equilibria that have at most discrete symmetry, one can apply the same procedure to the reduced vector field, where one sees a genuine equilibrium. The possible movement of eigenvalues that we focus on are illustrated by the following two cases:

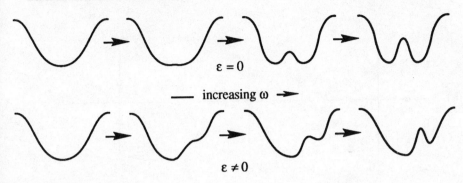

Figure 10.1.6: Potential for the centered and off-center ball in the hoop as the angular velocity increases.

1. *Steady-state bifurcation* The equilibrium has a zero eigenvalue of multiplicity two.

2. *1:1 resonance* The equilibrium has a pair of purely imaginary eigenvalues of multiplicity 2. (Without loss of generality, we may assume that these eigenvalues are $\pm i$.)

In Case 1, the kernel of the linearization is a two-dimensional symplectic subspace. As the bifurcation parameter is varied, generically the eigenvalues go from purely imaginary to real, or vice-versa.

In Case 2, the sum of the eigenspaces of the eigenvalues $\pm i$ can be written as the sum of two ω-orthogonal two-dimensional symplectic subspaces. This time, generically the eigenvalues go from purely imaginary into the right and left hand complex plane, or vice versa. In each of these cases we say that the eigenvalues *split*, (see Figure 10.2.1). The 1:1 resonance with splitting is often called the **Hamiltonian Hopf bifurcation** (see van der Meer [1985]).

In some applications the eigenvalues do not split at 0 or $\pm i$, but rather, they remain on the imaginary axis and *pass*, as in Figure 10.2.2.

Symmetry can influence the above generic behavior (see Golubitsky, Sheaffer and Stewart [1988]). Indeed, for certain symmetry groups (such as the circle group S^1), the passing of eigenvalues may be generic in a one parameter family. Galin [1982] shows that without symmetry, the generic situation is splitting and one would require three parameters to see passing.

In the steady state case, the dichotomy in eigenvalue movements can be understood using definiteness properties of the Hamiltonian, *i.e.*, by *energetics*, or *group theoretically*. The group theoretic approach was discussed in Golubitsky and Stewart [1987]. The energetics method has a

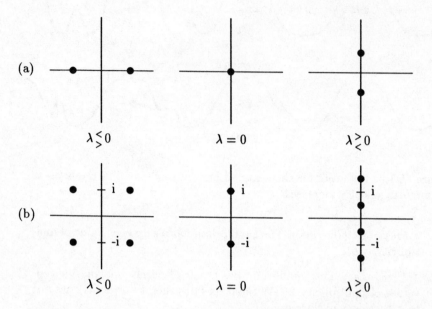

Figure 10.2.1: The *splitting* case; (a) for the steady state bifurcation, (b) for the 1:1 resonance.

complex history, going back to at least Cartan [1922] — a modern reference is Oh [1987]. For example, one can sometimes say that a S^1 symmetry forces the eigenvalues to stay on the imaginary axis, or one can say that eigenvalues must split because the second variation changes from positive definite to indefinite with just one eigenvalue crossing through zero.

Here are some simple observations about the method of energetics. Start with a linear (or linearized) Hamiltonian system and write it in the form $X_H = \mathbb{J}B$ where B is a symmetric matrix whose associated quadratic form is the Hamiltonian H and where $\mathbb{J} = \begin{bmatrix} 0 & 1 \\ -1 & 0 \end{bmatrix}$ is the Poisson tensor. If H is positive (or negative) definite, then the spectrum of X_H is on the imaginary axis — this follows because the system is necessarily stable, and the spectrum of X_H is symmetric under reflection in the two axis of the complex plane, so the spectrum must be confined to the imaginary axis. If H has an *odd* number of eigenvalues that are negative, then taking the determinant, we see that since $\mathbb{J}$ has determinant one, X_H has a negative determinant. Thus in this case, it must have at *least* one pair of real eigenvalues, and therefore be *linearly unstable*. If H has the standard form

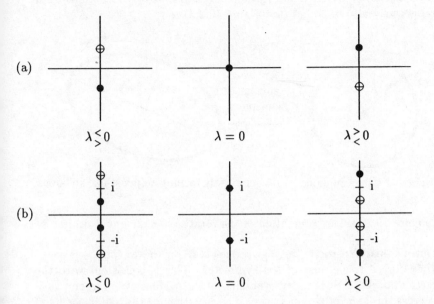

Figure 10.2.2: The *passing* case; (a) for the steady state bifurcation, (b) for the 1:1 resonance.

of kinetic plus potential energy, and the kinetic term is positive definite, and the potential energy has at least one negative eigenvalue, then again X_H has real eigenvalues, and so is *linearly unstable*. However, for gyroscopic systems, such as those that arise in reduction, the situation is not so simple, and deeper insight is needed.

An example relevant to the above remarks concerns bifurcations of relative equilibria of a rotating liquid drop: the system consists of the two dimensional Euler equations for an ideal fluid with a free boundary. A rigidly rotating circular drop is an equilibrium solution (in the spatially reduced equations). The energy-Casimir method shows stability, provided

$$\frac{\Omega^2}{12R^3\tau} < 1. \qquad (10.2.1)$$

Here Ω is the angular velocity of the circular drop, R is its radius and τ is the surface tension, a constant. As Ω increases and (10.2.1) is violated, the stability of the circular solution is lost and is picked up by elliptical-like solutions with $\mathbb{Z}_2 \times \mathbb{Z}_2$ symmetry. (The bifurcation is actually subcritical relative to Ω and is supercritical relative to the angular momentum.) This is proved in Lewis, Marsden and Ratiu [1987] and Lewis [1989], where other references may also be found (see Figure 10.2.3).

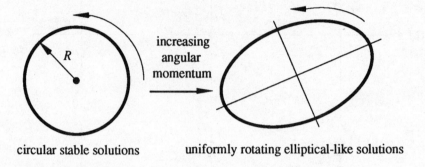

circular stable solutions   uniformly rotating elliptical-like solutions

Figure 10.2.3: The bifurcation of the rotating planar liquid drop.

During this transition, the eigenvalues stay on the imaginary axis — they are forced to because of the symmetry. This is consistent with the energetics approach since an *even* number of eigenvalues of the second variation cross through zero, namely two. The situation for the ball in the hoop and the liquid drop examples is presented in Figure 10.2.4.

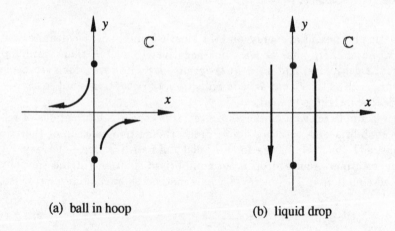

(a) ball in hoop   (b) liquid drop

Figure 10.2.4: Eigenvalue evolution for the liquid drop and the ball in the hoop.

Energetics or group theory alone is not sufficient to characterize the movement of eigenvalues in the 1:1 resonance. The work of Dellnitz, Melbourne and Marsden [1991] uses a combination of group theory and energetics that gives a particularly clean characterization of the splitting and passing cases. We summarize some relevant notation to explain these results.

Assume that the quadratic Hamiltonian is invariant under a compact Lie group Γ that preserves the symplectic structure. A Γ-invariant subspace V is called *absolutely* Γ-*irreducible* if the only linear mappings $V \to V$ that commute with the action of Γ are real multiples of the identity. An Γ-irreducible subspace that is not absolutely Γ-irreducible is called *nonabsolutely* Γ-*irreducible*. For example, the rotation group $SO(2)$ acting on the plane is nonabsolutely irreducible since any rotation commutes with this action, but nevertheless, the action is irreducible (has no nontrivial invariant subspaces). On the other hand, the rotation group $SO(3)$ acting in the usual way on three space is absolutely Γ-irreducible, as is easy to check.

Golubitsky and Stewart [1987] show that for steady-state bifurcation, generically the generalized zero eigenspace E_0 is either nonabsolutely Γ-irreducible or the direct sum of two isomorphic absolutely Γ-irreducible subspaces. These two possibilities correspond respectively to the splitting or passing of eigenvalues. In terms of energetics, the Hamiltonian changes from definite to indefinite in the splitting case, but remains definite in the passing case.

In the case of 1:1 resonance, generically the sum of the generalized eigenspaces of $\pm i$, $E_{\pm i}$, can be written as the sum of two symplectic ω-orthogonal subspaces U_1 and U_2, where each of the U_j is either nonabsolutely Γ-irreducible or the direct sum of two isomorphic absolutely Γ-irreducible subspaces.

The main result of Dellnitz, Melbourne and Marsden [1991] concerns the generic movement of eigenvalues in this situation. The most difficult cases are when U_1 and U_2 are isomorphic. In fact,

if U_1 and U_2 carry *distinct* representations of Γ then the resonance decouples and the eigenvalues move independently along the imaginary axis (*independent passing*).

To understand the cases where U_1 and U_2 are isomorphic, one uses the results of Montaldi, Roberts and Stewart [1989] on the relationship between the symmetry and the symplectic structure. At this stage it becomes necessary to distinguish between the two types of nonabsolutely Γ-irreducible representations: *complex* and *quaternionic*. Here, one uses the fact that for a Γ-irreducible representation, the space of linear mappings that commute with Γ is isomorphic to the reals, complexes, or to the quaternions. The real case corresponds to the absolutely irreducible case, and in the nonabsolutely irreducible case, one has either the complexes or the quaternions. If the U_j are isomorphic complex irreducibles, then in the terminology of Montaldi, Roberts and Stewart [1988], they are either of the *same type* or *dual*. When they are of the same type, one has a symplectic form that can be written as $\mathbb{J} = \begin{bmatrix} i & 0 \\ 0 & i \end{bmatrix}$ and in the case of duals, the symplectic form can be written as $\mathbb{J} = \begin{bmatrix} i & 0 \\ 0 & -i \end{bmatrix}$.

There are three cases:

1. Provided U_1 and U_2 are not complex irreducibles, generically the eigenvalues split and H is indefinite.

2. If U_1 and U_2 are complex of the same type, then generically the eigenvalues pass and H is indefinite.

3. In the case of complex duals the eigenvalues can generically pass or split and these possibilities correspond precisely to definiteness and indefiniteness of the quadratic form induced on $U_1 \oplus U_2$ by the linearization.

10.3   The One to One Resonance and Dual Pairs

An interesting discussion of the 1:1 resonance is given in Cushman and Rod [1982] (see also Marsden [1987]). Consider a family of Hamiltonians depending on a parameter λ near one of the form

$$H = \frac{1}{2}(q_1^2 + p_1^2) + \frac{\lambda}{2}(q_2^2 + p_2^2) + \text{ higher order terms.} \tag{10.3.1}$$

Notice that the quadratic part of this Hamiltonian has a $S^1 \times S^1$ symmetry acting separately on both factors and so we have, in the terminology of the preceding section, independent passing. Notice that H is definite in this case. The oscillators have the same frequency when $\lambda = 1$, corresponding to the 1:1 resonance. To analyze the dynamics of H, it is important to utilize a good geometric picture for the critical case when $\lambda = 1$ and we get the unperturbed Hamiltonian

$$H_0 = \frac{1}{2}(q_1^2 + p_1^2 + q_2^2 + p_2^2). \tag{10.3.2}$$

The energy level H_0 = constant is the three sphere $S^3 \subset \mathbb{R}^4$. If we think of H_0 as a function on $\mathbb{C}^2$ by letting

$$z_1 = q_1 + ip_1 \quad \text{and} \quad z_2 = q_2 + ip_2,$$

then $H_0 = (|z_1|^2 + |z_2|^2)/2$ and so H_0 is invariant under the action of $SU(2)$, the complex 2×2 unitary matrices of determinant one. The corresponding conserved quantities are

$$\begin{aligned}
W_1 &= 2(q_1 q_2 + p_1 p_2) \\
W_2 &= 2(q_2 p_1 - q_1 p_2) \\
W_3 &= q_1^2 + p_1^2 - q_2^2 - p_2^2
\end{aligned} \tag{10.3.3}$$

which comprise the components of a (momentum) map for the action of $SU(2)$ on $\mathbb{C}^2$:

$$\mathbf{J} : \mathbb{C}^2 \cong \mathbb{R}^4 \to su(2)^* \cong \mathbb{R}^3. \tag{10.3.4}$$

From the relation $4H_0^2 = W_1^2 + W_2^2 + W_3^2$, one finds that $\mathbf{J}$ restricted to S^3 gives a map

$$j : S^3 \to S^2. \tag{10.3.5}$$

The fibers $j^{-1}(\text{point})$ are circles and the dynamics of H_0 moves along these circles. The map j is the **Hopf fibration** which describes S^3 as a topologically nontrivial circle bundle over S^2. (The reduction of $\mathbb{R}^4$ by the action by the flow of H_0 is S^2.) Apparently the role of the Hopf fibration in mechanics was already known to Reeb around 1950.

With $P = \mathbb{C}^2$, we have a basic example of a dual pair determined by the above momentum maps (see Example 1 of §8.3 and the Example in §8.5). See Figure 10.3.1.

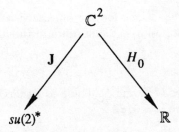

Figure 10.3.1: The dual pair of the harmonic oscillator.

Normal form theory allows one (up to finite order) to change coordinates by averaging over the S^1 action determined by the flow of H_0. In this way one gets a new Hamiltonian $\mathcal{H}$ from H that is S^1 invariant. Since we have a dual pair, such an $\mathcal{H}$ can be written as

$$\mathcal{H} = h \circ \mathbf{J}. \tag{10.3.6}$$

In other words, *a function invariant on one side collectivizes on the other*. In particular, since $\mathbf{J}$ is a Poisson map, *the dynamics of $\mathcal{H}$ can be reduced to dynamics on $su(2)^* \cong \mathbb{R}^3$ with the rigid body Lie-Poisson structure*. This proves one of the results of Cushman and Rod [1982]. This procedure can be of help in locating interesting bifurcations, as in David, Holm and Tratnik [1990].

The Hopf fibration occurs in a number of other interesting mechanical systems. One of these is the free rigid body. When doing reduction for the rigid body, we construct the reduced space $\mathbf{J}^{-1}(\mu)/G_\mu = \mathbf{J}^{-1}(\mu)/S^1$, which is the sphere S^2. Also, $\mathbf{J}^{-1}(\mu)$ is topologically the same as the rotation group $SO(3)$, which in turn is the same as $S^3/\mathbb{Z}_2$. Thus, the reduction map is a map of $SO(3)$ to S^2. Such a map is given explicitly by taking an orthogonal matrix A and mapping it to the vector on the sphere given by $A\mathbf{k}$, where $\mathbf{k}$ is the unit vector along the z-axis. This map that does the projection is in fact a restriction of a momentum map and, when composed with the map of $S^3 \cong SU(2)$ to $SO(3)$, is just the Hopf fibration again. Thus, not only does the Hopf fibration occur in the 1:1 resonance, *it occurs in the rigid body in a natural way as the reduction map from material to body representation!*

10.4   Bifurcations in the Double Spherical Pendulum

In §5.5 we wrote the equations for the linearized solutions of the double spherical pendulum at a relative equilibrium in the form

$$M\ddot{q} + S\dot{q} + \Lambda q = 0 \tag{10.4.1}$$

for certain $3{\times}3$ matrices M, S and Λ. These equations have the Hamiltonian form $\dot{F} = \{F, H\}$ where $p = M\dot{q}$,

$$H = \frac{1}{2}pM^{-1}P + \frac{1}{2}q\Lambda q \tag{10.4.2}$$

and

$$\{F, K\} = \frac{\partial F}{\partial q^i}\frac{\partial K}{\partial p_i} - \frac{\partial K}{\partial q^i}\frac{\partial F}{\partial p_i} - S_{ij}\frac{\partial F}{\partial p_i}\frac{\partial K}{\partial p_j} \tag{10.4.3}$$

i.e.,

$$\left. \begin{array}{l} \dot{q} = M^{-1}p \\ \dot{p} = -S\dot{q} - \Lambda q = -SM^{-1}p - \Lambda q. \end{array} \right\} \qquad (10.4.4)$$

The following is a standard useful observation:

Proposition 10.1 *The eigenvalues* λ *of the linear system (10.4.4) are given by the roots of*

$$\det[\lambda^2 M + \lambda S + \Lambda] = 0 \qquad (10.4.5)$$

Proof Let (u, v) be an eigenvector of (10.4.4) with eigenvalue λ; then

$$M^{-1}v = \lambda u \quad \text{and} \quad -SM^{-1}v - \Lambda u = \lambda v$$

i.e., $-S\lambda u - \Lambda u = \lambda^2 M u$, so u is an eigenvector of $\lambda^2 M + \lambda S + \Lambda$. ∎

For the double spherical pendulum, we call the eigenvalue γ (since λ is already used for something else in this example) and note that the polynomial

$$p(\gamma) = \det[\gamma^2 M + \gamma S + \Lambda] \qquad (10.4.6)$$

is cubic in γ^2, as it must be, consistent with the symmetry of the spectrum of Hamiltonian systems. This polynomial can be analyzed for specific system parameter values. In particular, for $r = 1$ and $\overline{m} = 2$, *one finds a Hamiltonian Hopf bifurcation along the cowboy branch as we go up the branch in Figure 4.3.1 with increasing* λ *starting at* $\alpha = -\sqrt{2}$.

The situation before the bifurcation (for smaller λ and μ), is one where the energetics method and the spectral method disagree in their conclusions about stability. The situation will be resolved in §10.7.

Perhaps more interesting is the fact that for certain system parameters, the Hamiltonian Hopf point can converge to the straight down *singular*(!) state with $\lambda = 0 = \mu$. Here, (10.4.1) does not make sense, and must be *regularized*. After this is done, one finds (with still two parameters left) that one has a system in which both passing and splitting can generically occur. One can hope that the ideas of §10.2 with the inclusion of an *antisymplectic* reversibility type of symmetry will help to explain this observed phenomenon. We refer to Dellnitz, Marsden, Melbourne and Scheurle [1992] for further details.

10.5 Continuous Symmetry Groups and Solution Space Singularities

Recall that *singular points of* **J** *are points with symmetry*. This turns out to be a profound observation with far reaching implications. The level sets of

J typically have *quadratic* singularities at its singular (= symmetric) points, as was shown by Arms, Marsden and Moncrief [1981]. In the abelian case, the images of these symmetric points are the vertices, edges and faces of the convex polyhedron $\mathbf{J}(P)$ in the Atiyah-Guillemin-Sternberg convexity theory. (See Atiyah [1982] and Guillemin and Sternberg [1984].) As one leaves this singular point, heading for a generic one with no singularities, one passes through the lattice of isotropy subgroups of G. Arms, Marsden and Moncrief [1981] describe how these symmetry groups break.

These ideas apply in a remarkable way to solution spaces of relativistic field theories, such as Einstein's equations of general relativity and the Yang-Mills equations on space time. Here the theories have symmetry groups and, appropriately interpreted, corresponding momentum maps. The relativistic field equations split into two parts — Hamiltonian hyperbolic evolution equations and elliptic constraint equations. The solution space structure is determined by the elliptic constraint equations, which in turn say exactly that the momentum map vanishes.

A fairly long story of both geometry and analysis is needed to establish this, but the result can be simply stated: the *solution space has a quadratic singularity precisely at those field points that have symmetry*. For further details, see Fischer, Marsden and Moncrief [1981] and Arms, Marsden and Moncrief [1982].

While these results were motivated by perturbation theory of classical solutions (gravitational waves as solutions of the linearized Einstein equations etc.), there is some evidence that these singularities have quantum implications. For example, there appears to be evidence that in the Yang-Mills case, wave functions tend to concentrate near singular points (see, for example, Emerich and Römer [1990]).

For bifurcation theory, as we have indicated in the preceeding sections, a start has been made on how to tackle the problem when there is a continuous isotropy group and some examples have been worked out. One of these is the bifurcations in a rotating liquid drop, already mentioned above, where the isotropy group is the whole symmetry group S^1. Here the problems with the singular structure of the momentum map are obviated by working with the spatially reduced system, and the energy-Casimir method. Here, the symmetry is dealt with by directly factoring it out "by hand", using appropriately defined polar coordinates. Another case that is dealt with is the heavy top in Lewis, Ratiu, Simo and Marsden [1992]. Here, bifurcations emanate from the upright position, which has a nondiscrete symmetry group S^1. That paper, and Lewis [1991] indicate how a general theory of stability might go in the presence of general isotropy groups. Presumably the general bifurcation theory will follow.

10.6   The Poincaré-Melnikov Method

To begin with a simple example, consider the equation of a forced pendulum

$$\ddot{\phi} + \sin\phi = \epsilon\cos\omega t. \tag{10.6.1}$$

Here ω is a constant angular forcing frequency, and ϵ is a small parameter. For $\epsilon = 0$ this has the phase portrait of a simple pendulum. For ϵ small but non-zero, (10.6.1) possesses no analytic integrals of the motion. In fact, it possesses transversal intersecting stable and unstable manifolds (separatrices); that is, the Poincaré maps $P_{t_0} : \mathbb{R}^2 \to \mathbb{R}^2$ that advance solutions by one period $T = 2\pi/\omega$ starting at time t_0 possess transversal homoclinic points. This type of dynamic behavior has several consequences, besides precluding the existence of analytic integrals, that lead one to use the term "chaotic". For example, Equation (10.6.1) has infinitely many periodic solutions of arbitrarily high period. Using the shadowing lemma, one sees that given any bi-infinite sequence of zeros and ones (for example, use the binary expansion of e or π), there exists a corresponding solution of (10.6.1) that successively crosses the plane $\phi = 0$ (the pendulum's vertically downward configuration) with $\phi > 0$ corresponding to a zero and $\phi < 0$ corresponding to a one. The origin of this chaos on an intuitive level lies in the motion of the pendulum near its unperturbed homoclinic orbit — the orbit that does one revolution in infinite time. Near the top of its motion (where $\phi = \pm\pi$) small nudges from the forcing term can cause the pendulum to fall to the left or right in a temporally complex way.

The Poincaré-Melnikov method is as follows: First, write the dynamical equation to be studied in abstract form as

$$\dot{x} = X_0(x) + \epsilon X_1(x, t) \tag{10.6.2}$$

where $x \in \mathbb{R}^2$, X_0 is a Hamiltonian vector field with energy H_0, X_1 is Hamiltonian with energy a T-periodic function H_1. Assume that X_0 has a homoclinic orbit $\bar{x}(t)$ so $\bar{x}(t) \to x_0$, a hyperbolic saddle point, as $t \to \pm\infty$. Second, compute the *Poincaré-Melnikov function* defined by

$$M(t_0) = \int_{-\infty}^{\infty} \{H_0, H_1\}(\bar{x}(t - t_0), t)dt \tag{10.6.3}$$

where $\{\,,\,\}$ denotes the Poisson bracket.

Theorem 10.1 *Poincaré-Melnikov If $M(t_0)$ has simple zeros as a function of t_0, then (10.6.2) has, for sufficiently small ϵ, homoclinic chaos in the sense of transversal intersecting separatrices.*

We shall give a proof of this result below and in the course of the proof, we shall clarify what it means to have transverse separatrices.

To apply this method to equation (10.6.1), let $x = (\phi, \dot\phi)$ so (10.6.1) becomes

$$\frac{d}{dt} \begin{bmatrix} \phi \\ \dot\phi \end{bmatrix} = \begin{bmatrix} \dot\phi \\ -\sin\phi \end{bmatrix} + \epsilon \begin{bmatrix} 0 \\ \cos\omega t \end{bmatrix}. \tag{10.6.4}$$

The homoclinic orbits for $\epsilon = 0$ are computed to be given by

$$\bar{x}(t) = \begin{bmatrix} \phi(t) \\ \dot\phi(t) \end{bmatrix} = \begin{bmatrix} \pm 2 \tan^{-1}(\sinh t) \\ \pm 2\,\mathrm{sech}\,t \end{bmatrix}$$

and one has

$$H_0(\phi, \dot\phi) = \frac{1}{2}\dot\phi^2 - \cos\phi, \quad \text{and} \quad H_1(\phi, \dot\phi, t) = \phi\cos\omega t.$$

Hence (10.6.3) gives

$$\begin{aligned}
M(t_0) &= \int_{-\infty}^{\infty} \left(\frac{\partial H_0}{\partial\phi}\frac{\partial H_1}{\partial\dot\phi} - \frac{\partial H_0}{\partial\dot\phi}\frac{\partial H_1}{\partial\phi} \right)(\bar{x}(t - t_0), t)dt \\
&= -\int_{-\infty}^{\infty} \dot\phi(t - t_0)\cos\omega t\,dt \\
&= \mp\int_{-\infty}^{\infty} [2\,\mathrm{sech}\,(t - t_0)\cos\omega t]dt. \tag{10.6.5}
\end{aligned}$$

Changing variables and using the fact that sech is even and sin is odd, we get

$$M(t_0) = \mp 2 \left(\int_{-\infty}^{\infty} \mathrm{sech}\,t\cos\omega t\,dt \right) \cos(\omega t_0).$$

The integral is evaluated by residues:

$$M(t_0) = \mp 2\pi\,\mathrm{sech}\left(\frac{\pi\omega}{2} \right) \cos(\omega t_0),$$

which clearly has simple zeros. Thus, this equation has chaos for ϵ small enough.

Now we turn to a proof of the Poincaré-Melnikov theorem. There are two convenient ways of visualizing the dynamics of (10.6.2). Introduce the Poincaré map $P_\epsilon^s : \mathbb{R}^2 \to \mathbb{R}^2$, which is the time T map for (10.6.2) starting at time s. For $\epsilon = 0$, the point x_0 and the homoclinic orbit are invariant under P_0^s, which is independent of s. The hyperbolic saddle x_0 persists as a nearby family of saddles x_ϵ for $\epsilon > 0$, small, and we are interested in whether or not the stable and unstable manifolds of the point x_ϵ for the map P_ϵ^s intersect transversally (if this holds for one s, it holds for all s). If so, we say (10.6.2) **admits horseshoes for $\epsilon > 0$**.

The second way to study (10.6.2) is to look directly at the **suspended system** on $\mathbb{R}^2 \times S^1$:

$$\begin{aligned}
\dot{x} &= X_0(x) + \epsilon X_1(x, \theta), \\
\dot{\theta} &= 1.
\end{aligned} \qquad (10.6.6)$$

From this point of view, the curve

$$\gamma_0(t) = (x_0, t)$$

is a periodic orbit for (10.6.2), whose stable and unstable manifolds $W_0^s(\gamma_0)$ and $W_0^u(\gamma_0)$ are coincident. For $\epsilon > 0$ the hyperbolic closed orbit γ_0 perturbs to a nearby hyperbolic closed orbit which has stable and unstable manifolds $W_\epsilon^s(\gamma_\epsilon)$ and $W_\epsilon^u(\gamma_\epsilon)$. If $W_\epsilon^s(\gamma_\epsilon)$ and $W_\epsilon^u(\gamma_\epsilon)$ intersect transversally, we again say that (10.6.2) **admits horseshoes**. These two definitions of admitting horseshoes are equivalent.

We use the energy function H_0 to measure the first order movement of $W_\epsilon^s(\gamma_\epsilon)$ at $\bar{x}(0)$ at time t_0 as ϵ is varied. Note that points of $\bar{x}(t)$ are regular points for H_0 since H_0 is constant on $\bar{x}(t)$ and $\bar{x}(0)$ is not a fixed point. Thus, the values of H_0 can be used to measure the distance from the homoclinic orbit. If $(x_\epsilon^s(t, t_0), t)$ is the curve on $W_\epsilon^s(\gamma_\epsilon)$ that is an integral curve of the suspended system in xt-space, and has an initial condition $x^s(t_0, t_0)$ which is the perturbation of $W_0^s(\gamma_0) \cap \{$ the plane $t = t_0\}$ in the normal direction to the homoclinic orbit, then $H_0(x_\epsilon^s(t_0, t_0))$ measures the normal distance. But

$$H_0(x_\epsilon^s(T, t_0)) - H_0(x_\epsilon^s(t_0, t_0)) = \int_{t_0}^T \frac{d}{dt} H_0(x_\epsilon^s(t, t_0)) dt, \qquad (10.6.7)$$

and so

$$H_0(x_\epsilon^s(T, t_0)) - H_0(x_\epsilon^s(t_0, t_0)) = \int_{t_0}^T \{H_0, H_0 + \epsilon H_1\}(x_\epsilon^s(t, t_0), t) dt. \quad (10.6.8)$$

Since $x_\epsilon^s(T, t_0)$ is ϵ-close to $\bar{x}(t - t_0)$ (uniformly as $T \to +\infty$),

$$\mathbf{d}(H_0 + \epsilon H^1)(x_\epsilon^s(t, t_0), t) \to 0$$

exponentially as $t \to +\infty$, and $\{H_0, H_0\} = 0$, so (10.6.8) becomes

$$H_0(x_\epsilon^s(T, t_0)) - H_0(x_\epsilon^s(t_0, t_0)) = \epsilon \int_{t_0}^T \{H_0, H_1\}(\bar{x}(t - t_0, t)) dt + O(\epsilon^2).$$
$$(10.6.9)$$

Similarly,

$$H_0(x_\epsilon^u(t_0, t_0)) - H_0(x_\epsilon^u(-S, t_0)) = \epsilon \int_{-S}^{t_0} \{H_0, H_1\}(\bar{x}(t - t_0, t))dt + O(\epsilon^2).$$
(10.6.10)

Since $x_\epsilon^s(T, t_0) \to \gamma_\epsilon$, a periodic orbit for the perturbed system as $T \to +\infty$, we can choose T and S such that $H_0(x_\epsilon^s(T, t_0)) - H_0(x_\epsilon^u(-S, t_0)) \to 0$ as $T, S \to \infty$. Thus, adding (10.6.9) and (10.6.10), and letting $T, S \to \infty$, we get

$$H_0(x_\epsilon^u(t_0, t_0)) - H_0(x_\epsilon^s(t_0, t_0)) = \epsilon \int_{-\infty}^{\infty} \{H_0, H_1\}(\bar{x}(t - t_0, t))dt + O(\epsilon^2).$$
(10.6.11)

It follows that if $M(t_0)$ has a simple zero at time t_0, then $x_\epsilon^u(t_0, t_0)$ and $x_\epsilon^s(t_0, t_0)$ intersect transversally near the point $\bar{x}(0)$ at time t_0. (Since $dH_0 \to 0$ exponentially at the saddle points, the integrals involved in this criterion are automatically convergent.) ∎

We now describe a few of the extensions and applications of this technique. The literature in this area is growing very quickly and we make no claim to be comprehensive (the reader can track down many additional references by consulting Wiggins [1988] and the references cited below).

If in (10.6.2), X_0 only is a Hamiltonian vector field, the same conclusion holds if (10.6.3) is replaced by

$$M(t_0) = \int_{-\infty}^{\infty} (X_0 \times X_1)(\bar{x}(t - t_0), t)dt, \tag{10.6.12}$$

where $X_0 \times X_1$ is the (scalar) cross product for planar vector fields. In fact, X_0 need not even be Hamiltonian if an area expansion factor is inserted. For example, (10.6.12) applies to the forced damped Duffing equation

$$\ddot{u} - \beta u + \alpha u^3 = \epsilon(\gamma \cos \omega t - x\dot{u}). \tag{10.6.13}$$

Here the homoclinic orbits are given by

$$\ddot{u}(t) = \pm\sqrt{\frac{\beta}{\alpha}} \operatorname{sech}(\beta^{\frac{1}{2}}t) \tag{10.6.14}$$

and (10.6.12) becomes, after a residue calculation,

$$M(t_0) = 2\gamma\pi\omega\sqrt{\frac{2}{\alpha}} \operatorname{sech}\left(\frac{\pi\omega}{2\sqrt{\beta}}\right) \sin(\omega t_0) + \frac{4\delta\beta^{\frac{3}{2}}}{3\alpha} \tag{10.6.15}$$

so one has simple zeros and hence chaos of the horseshoe type if

$$\frac{\gamma}{\delta} > \frac{\sqrt{2}\beta^{\frac{3}{2}}}{3\omega\sqrt{\alpha}} \cosh\left(\frac{\pi\omega}{2\sqrt{\beta}}\right) \tag{10.6.16}$$

and ϵ is small.

Another interesting example, due to Montgomery, concerns the equations for superfluid ^{3}He. These are the Leggett equations and we shall confine ourselves to the A phase for simplicity. The equations are

$$\dot{s} = -\frac{1}{2}\left(\frac{\chi\Omega^2}{\gamma^2}\right)\sin 2\theta \quad \text{and} \quad \dot{\theta} = \left(\frac{\gamma^2}{\chi}\right)s - \epsilon(\gamma B \sin\omega t + \frac{1}{2}\Gamma\sin 2\theta). \tag{10.6.17}$$

Here s is the spin, θ an angle (describing the order parameter) and $\gamma, \chi, \ldots$ are physical constants. The homoclinic orbits for $\epsilon = 0$ are given by

$$\bar{\theta}_{\pm} = 2\tan^{-1}(e^{\pm\Omega t}) - \pi/2 \quad \text{and} \quad \bar{s}_{\pm} = \pm 2\frac{\Omega e^{\pm 2\Omega t}}{1 + e^{\pm 2\Omega t}}. \tag{10.6.18}$$

One calculates, after substituting (10.6.18) and (10.6.19) in (10.6.12) that

$$M_{\pm}(t_0) = \mp \frac{\pi\chi\omega B}{8\gamma} \operatorname{sech}\left(\frac{\omega\pi}{2\Omega}\right)\cos\omega t - \frac{2}{3}\frac{\chi}{\gamma^2}\Omega\Gamma \tag{10.6.19}$$

so that (10.6.17) has chaos in the sense of horseshoes if

$$\frac{\gamma B}{\Gamma} > \frac{16}{3\pi}\frac{\Omega}{\omega}\cosh\left(\frac{\pi\omega}{2\Omega}\right) \tag{10.6.20}$$

and if ϵ is small.

A version of the Poincaré-Melnikov theorem applicable to PDE's (due to Holmes and Marsden [1981]). One basically still uses formula (10.6.12) where $X_0 \times X_1$ is replaced by the symplectic pairing between X_0 and X_1. However, there are two new difficulties in addition to standard technical analytic problems that arise with PDE's. The first is that there is a serious problem with resonances. These can be dealt with using the aid of damping — the undamped case would need an infinite dimensional version of Arnold diffusion. Secondly, the problem is not reducible to two dimensions; the horseshoe involves all the modes. Indeed, the higher modes do seem to be involved in the physical buckling processes for the beam model discussed next.

A PDE model for a buckled forced beam is

$$\ddot{w} + w'''' + \Gamma w' - K\left(\int_0^1 [w']^2 dz\right)w'' = \epsilon(f\cos\omega t - \delta\dot{w}) \tag{10.6.21}$$

where $w(z,t), 0 \leq z \leq 1$ describes the deflection of the beam, $\dot{} = \partial/\partial t, ' = \partial/\partial z$ and $\Gamma, K, \ldots$ are physical constants. For this case, the theory shows that if

1. $\pi^2 < \Gamma < 4\rho^3$ (first mode is buckled)

2. $j^2\pi^2(j^2\pi^2 - \Gamma) \neq \omega^2, j = 2, 3, \ldots$ (resonance condition)

3. $\dfrac{f}{\delta} > \dfrac{\pi(\Gamma - \pi^2)}{2\omega\sqrt{K}} \cosh\left(\dfrac{\omega}{2\sqrt{\Gamma - \omega^2}}\right)$ (transversal zeros for $M(t_0)$)

4. $\delta > 0$

and ϵ is small, then (10.6.21) has horseshoes. Experiments (see Moon and Holmes [1979]) showing chaos in a forced buckled beam provided the motivation which lead to the study of (10.6.21).

This kind of result can also be used for a study of chaos in a van der Waal fluid (Slemrod and Marsden [1983]) and there is a growing literature using these methods for forced and damped soliton equations. For example, in the damped, forced Sine-Gordon equation one has chaotic transitions between breathers and kink-antikink pairs and in the Benjamin-Ono equation one can have chaotic transitions between solutions with different numbers of poles.

For Hamiltonian systems with two degrees of freedom, Holmes and Marsden [1982a] show how the Melnikov method may be used to prove the existence of horseshoes on energy surfaces in nearly integrable systems. The class of systems studied have a Hamiltonian of the form

$$H(q, p, \theta, I) = F(q, p) + G(I) + \epsilon H_1(q, p, \theta, I) + O(\epsilon^2) \qquad (10.6.22)$$

where (θ, I) are action-angle coordinates for the oscillator $G; G(0) = 0, G' > 0$. It is assumed that F has a homoclinic orbit $\bar{x}(t) = (\bar{q}(t), \bar{p}(t))$ and that

$$M(t_0) = \int_{-\infty}^{\infty} \{F, H_1\}dt, \qquad (10.6.23)$$

the integral taken along $(\bar{x}(t - t_0), \Omega t, I)$, has simple zeros. Then (10.6.22) has horseshoes on energy surfaces near the surface corresponding to the homoclinic orbit and small I; the horseshoes are taken relative to a Poincaré map strobed to the oscillator G. The paper Holmes and Marsden [1982a] also studies the effect of positive and negative damping. These results are related to those for forced one degree of freedom systems since one can often reduce a two degrees of freedom Hamiltonian system to a one degree of freedom forced system.

For some systems in which the variables do not split as in (10.6.22), such as a nearly symmetric heavy top, one can exploit symmetry of the system and make use of reduction ideas. The general theory for this is given in Holmes and Marsden [1983] and was applied to show the existence of horseshoes in the nearly symmetric heavy top; see also some closely related results of Ziglin [1981].

The Poincaré-Melnikov theory has been used by Ziglin [1980b] in vortex dynamics, for example to give a proof of the non-integrability of the restricted four vortex problem. There have also been recent applications to the dynamics of general relativity showing the existence of horseshoes in Bianchi IX models. See Oh et al. [1989] for applications to the dynamics of coupled planar rigid bodies and to David, Holm and Tratnik [1990] to the study of polarization laser dynamics.

Arnold [1964] extended the Poincaré-Melnikov theory to systems with several degrees of freedom. In this case the transverse homoclinic manifolds are based on KAM tori and allow the possibility of chaotic drift from one torus to another. This drift, now known as *Arnold diffusion* is a basic ingredient in the study of chaos in Hamiltonian systems (see for instance, Chirikov [1979] and Lichtenberg and Lieberman [1983] and references therein). Instead of a single Melnikov function, one now has a *Melnikov vector* given schematically by

$$\vec{M} = \begin{pmatrix} \int_{-\infty}^{\infty} \{H_0, H_1\} dt \\ \int_{-\infty}^{\infty} \{I_k, H_1\} dt \end{pmatrix} \tag{10.6.24}$$

where I_k are integrals for the unperturbed (completely integrable) system and where $\vec{M}$ depends on t_0 and on angles conjugate to $I_1, \ldots, I_n$. One requires $\vec{M}$ to have transversal zeros in the vector sense. This result was given by Arnold for forced systems and was extended to the autonomous case by Holmes and Marsden [1982b, 1983]; see also Robinson [1988]. These results apply to systems such as pendulum coupled to several oscillators and the many vortex problems. It has also been used in power systems by Salam, Marsden and Varaiya [1983], building on the horseshoe case treated by Kopell and Washburn [1982]. There have been a number of other directions of research on these techniques. For example, Grundler developed a multidimensional version applicable to the spherical pendulum and Greenspan and Holmes showed how it can be used to study subharmonic bifurcations. See Wiggins [1988] references and for more information.

In Poincaré's celebrated memoir [1890] on the three-body problem, he introduced the mechanism of transversal intersection of separatrices which obstructs the integrability of the equations and the attendant convergence of series expansions for the solutions. This idea was subsequently developed by Birkhoff and Smale using the horseshoe construction to describe the resulting chaotic dynamics. However, in the region of phase space studied by Poincaré, it has never been proved (except in some generic sense that is not easy to interpret in specific cases) that the equations really are nonintegrable. In fact Poincaré himself traced the difficulty to the presence of terms in the separatrix splitting which are *exponentially small*. A crucial component of the measure of the splitting is given by the following formula of Poincaré [1890, p. 223]:

$$J = \frac{-8\pi i}{\exp\left(\frac{\pi}{\sqrt{2\mu}}\right) + \exp\left(-\frac{\pi}{\sqrt{2\mu}}\right)}$$

which is exponentially small (or beyond all orders) in μ. Poincaré was well aware of the difficulties that this exponentially small behavior causes; on p. 224 of his article, he states: "En d'autres termes, si on regarde μ comme un infiniment petit du premier ordre, la distance BB' sans être nulle, est un infiniment petit d'ordre infini. C'est ainsi que la fonction $e^{-1/\mu}$ est un infiniment petit d'ordre infini sans ètre nulle ... Dans l'example particulier que nous avons traité plus haut, la distance BB' est du mème ordre de grandeur que l'integral J, c'est à dire que $\exp(-\pi/\sqrt{2\mu})$."

This is a serious difficulty that arises when one uses the Melnikov method near an elliptic fixed point in a Hamiltonian system or in bifurcation problems giving birth to homoclinic orbits. The difficulty is related to those described by Poincaré (see Sanders [1982]). Near elliptic points, one sees homoclinic orbits in normal forms and after a temporal rescaling, this leads to a rapidly oscillatory perturbation that is modelled by the following variation of (10.6.1):

$$\ddot{\phi} + \sin\phi = \epsilon \cos\left(\frac{\omega t}{\epsilon}\right). \qquad (10.6.25)$$

If one formally computes $M(t_0)$ one finds from (10.6.3):

$$M(t_0, \epsilon) = \pm 2\pi \operatorname{sech}\left(\frac{\pi\omega}{2\epsilon}\right) \cos\left(\frac{\omega t_0}{\epsilon}\right). \qquad (10.6.26)$$

While this has simple zeros, the proof of the Poincaré-Melnikov theorem is no longer valid since $M(t_0, \epsilon)$ is now of order $e^{-\pi/2\epsilon}$ and the error analysis in the proof only gives errors of order ϵ^2. In fact, no expansion in powers of ϵ can detect exponentially small terms like $e^{-\pi/2\epsilon}$.

Holmes, Marsden and Scheurle [1989], Scheurle [1989], and Scheurle, Holmes and Marsden [1991] show, amongst other things, that if (10.6.25) has chaos, it is exponentially small in ϵ. Not only that, examples show how truly subtle this situation is, and one has to be extremely careful with hidden assumptions in the literature (like a priori hypotheses about the order of magnitude of the splitting bein of the form $\epsilon^k e^{-c/\epsilon}$ for some k and ϵ) that can be false in some examples. To get such estimates, the extension of the system to complex time plays a crucial role. One can hope that if sharp enough such results for (10.6.25) can really be proven, then it may be possible to return to Poincaré's 1890 work and complete the arguments he left unfinished.

To illustrate how exponentially small phenomena enter bifurcation problems, consider the problem of a Hamiltonian saddle node bifurcation

$$\ddot{x} + \mu x + x^2 = 0 \tag{10.6.27}$$

with the addition of higher order terms and forcing:

$$\ddot{x} + \mu x + x^2 + \text{h.o.t.} = \delta f(t). \tag{10.6.28}$$

The phase portrait of (10.6.27) is shown in Figure 10.6.1.

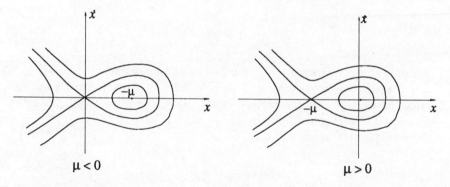

Figure 10.6.1: The evolution of the phase portrait of (10.6.27) as μ increases.

The system (10.6.27) is Hamiltonian with

$$H(x,p) = \frac{1}{2}p^2 + \frac{1}{2}\mu x^2 + \frac{1}{3}x^3. \tag{10.6.29}$$

Let us first consider the system without higher order terms:

$$\ddot{x} + \mu x + x^2 = \delta f(t). \tag{10.6.30}$$

To study it, we rescale to blow up the singularity:

$$x(t) = \lambda \xi(t) \tag{10.6.31}$$

where $\lambda = \|\mu\|$ and $t = t\sqrt{\lambda}$. We get

$$\ddot{\xi} - \xi + \xi^2 = \frac{\delta}{\mu^2} f\left(\frac{\tau}{\sqrt{-\mu}}\right), \mu < 0,$$

$$\ddot{\xi} - \xi + \xi^2 = \frac{\delta}{\mu^2} f\left(\frac{\tau}{\sqrt{\mu}}\right), \mu < 0. \tag{10.6.32}$$

The exponentially small estimates of Holmes, Marsden and Scheurle [1989] apply to (10.6.32). One gets upper and lower estimates in certain algebraic sectors of the (δ, μ) plane.

Now we consider

$$\ddot{x} + \mu x + x^2 + x^3 = \delta f(t). \tag{10.6.33}$$

With $\delta = 0$, there are equilibria at

$$x = 0, -r, \quad \text{or} \quad -\frac{\mu}{r} \quad \text{and} \quad \dot{x} = 0, \tag{10.6.34}$$

where

$$r = \frac{1 + \sqrt{1 - 4\mu}}{2}, \tag{10.6.35}$$

which is approximately 1 when $\mu \approx 0$. The phase portrait of (10.6.33) with $\delta = 0$ and $\mu = -\frac{1}{2}$ is shown in Figure 10.6.2. As μ passes through 0, the small lobe undergoes the same bifurcation as in Figure 10.6.1, with the large lobe changing only slightly.

Again we rescale by (10.6.35) to give

$$\ddot{\xi} - \xi + \xi^2 - \mu\xi^3 = \frac{\delta}{\mu^2} f\left(\frac{\tau}{\sqrt{-\mu}}\right), \mu < 0,$$

$$\ddot{\xi} - \xi + \xi^2 + \mu\xi^3 = \frac{\delta}{\mu^2} f\left(\frac{\tau}{\sqrt{\mu}}\right), \mu < 0. \tag{10.6.36}$$

Notice that for $\delta = 0$, the phase portrait is μ-dependent. The homoclinic orbit surrounding the small lobe for $\mu < 0$ is given explicitly in terms of ξ by

$$\xi(\tau) = \frac{4e^{\tau}}{\left(e^{\tau} + \frac{2}{3}\right)^2 - 2\mu}, \tag{10.6.37}$$

which is μ-dependent. An interesting technicality is that without the cubic term, we get μ-independent double poles at $\tau = \pm i\pi + \log 2 - \log 3$ in the

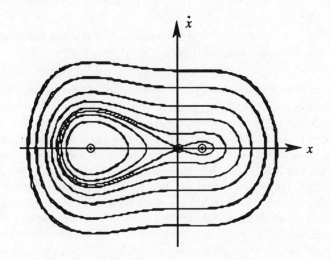

Figure 10.6.2: The phase portrait of 10.6.31 with $\delta = 0$.

complex τ-plane, while (10.6.37) has a pair of simple poles that splits these double poles to the pairs of simple poles at

$$\tau = \pm i\pi + \log\left(\frac{2}{3} \pm i\sqrt{2\lambda}\right) \tag{10.6.38}$$

where again $\lambda = \|\mu\|$. (There is no particular significance to the real part, such as $\log 2 - \log 3$ in the case of no cubic term; this can always be gotten rid of by a shift in the base point $\xi(0)$.)

If a quartic term x^4 is added, these pairs of simple poles will split into quartets of branch points and so on. Thus, while the analysis of higher order terms has this interesting μ-dependence, it seems that the basic exponential part of the estimates,

$$\exp\left(-\frac{\pi}{\sqrt{\|\mu\|}}\right), \tag{10.6.39}$$

remains intact.

10.7   The Role of Dissipation

If z_e is an equilibrium point of a Hamiltonian vector field X_H, then there are two methodologies for studying stability, as we already saw in the introductory chapter.

a *Energetics* — determine if $\delta^2 H(z_e) = Q$ is a definite quadratic form (Lagrange-Dirichlet).

b *Spectral methods* — determine if the spectrum of the linearized operator $\mathbf{D}X_H(z_e) = L$ is on the imaginary axis.

The energetics method can, via reduction, be applied to relative equilibria too and is the basis of the energy-momentum method that we studied in Chapter 5.

For general (not necessarily Hamiltonian) vector fields, the classical Liapunov theorem states that if the spectrum of the linearized equations lies strictly in the left half plane, then the equilibrium is stable and even asymptotically stable (trajectories starting close to the equilibrium converge to it exponentially as $t \to \infty$). Also, if any eigenvalue is in the strict right half plane, the equilibrium is unstable. This result, however, cannot apply to the purely Hamiltonian case since the spectrum of L is invariant under reflection in the real and imaginary coordinate axes. Thus, the only possible spectral configuration for a stable point of a Hamiltonian system is if the spectrum is *on* the imaginary axis.

The relation between **a** and **b** is, in general, complicated, but one can make some useful elementary observations.

Remarks

1 Definiteness of Q implies spectral stability (*i.e.*, the spectrum of L is on the imaginary axis). This is because spectral instability implies (linear and nonlinear) instability, while definiteness of Q implies stability.

2 Spectral stability need not imply stability, even linear stability. This is shown by the unstable linear system $\dot{q} = p, \dot{p} = 0$ with a pair of eigenvalues at zero.

3 If Q has odd index (an odd number of negative eigenvalues), then L has a real positive eigenvalue; see Oh [1987]. Indeed, in canonical coordinates, and identifying Q with its corresponding matrix, we have $L = \mathbb{J}Q$. Thus, $\det L = \det Q$ is negative. Since $\det L$ is the product of the eigenvalues of L and they come in conjugate pairs, there must be at least one real pair of eigenvalues, and in fact an odd number of positive real eigenvalues.

4 If $P = T^*Q$ with the standard symplectic structure (no magnetic terms) and if H is of the form kinetic plus potential so that an equilibrium has the form $(q_e, 0)$, and if $\delta^2 V(q_e)$ has negative index, then again L must have real eigenvalues. This is because one can diagonalize $\delta^2 V(q_e)$ with respect to the kinetic energy inner product, in which case the eigenvalues are evident.

◆

To get more interesting effects than covered by the above remarks, we consider *gyroscopic systems*; *i.e.*, linear systems of the form

$$M\ddot{q} + S\dot{q} + \Lambda q = 0 \qquad (10.7.1)$$

where M is a positive definite symmetric $n \times n$ matrix, S is skew, and Λ is symmetric. The term with S is the gyroscopic, or magnetic term. As we observed earlier, this system is verified to be Hamiltonian with $p = M\dot{q}$, energy function

$$H(q,p) = \frac{1}{2}pM^{-1}p + \frac{1}{2}q\Lambda q \qquad (10.7.2)$$

and the bracket

$$\{F, K\} = \frac{\partial F}{\partial q^i}\frac{\partial K}{\partial p_i} - \frac{\partial K}{\partial q^i}\frac{\partial F}{\partial p_i} - S_{ij}\frac{\partial F}{\partial p_i}\frac{\partial K}{\partial p_j}. \qquad (10.7.3)$$

If the index of V is even (see Remark **3**) one can get situations where $\delta^2 H$ is indefinite and yet spectrally stable. Roughly, this is a situation that is capable of undergoing a Hamiltonian Hopf bifurcation, so includes examples like the "cowboy" solution for the double spherical pendulum and certain regimes of the heavy top.

Theorem 10.2 *Dissipation induced instabilities Under the above conditions, if we modify* (10.7.1) *to*

$$M\ddot{q} + (S + \epsilon R)\dot{q} + \Lambda q = 0 \qquad (10.7.4)$$

for small $\epsilon > 0$ and R symmetric and positive definite, then the perturbed linearized equations $\dot{z} = L_\epsilon z$ are spectrally unstable, i.e., at least one pair of eigenvalues of L_ϵ is in the right half plane.

This result, due to Bloch, Krishnaprasad, Marsden and Ratiu [1992] builds on basic work of Chetaev [1961] and Hahn [1967]. The argument proceeds in two steps.

Step 1 Construct the *Chetaev functional*

$$W(q,p) = H(q,p) + \beta p \cdot (\Lambda q) \qquad (10.7.5)$$

for small β.

This function has the beautiful property that for β small enough, W has the same index as H, yet $\dot{W}$ is negative definite, where the overdot is taken in the dynamics of (10.7.4). This alone is enough to prove Liapunov instability, as is seen by studying the equation

$$W(q(T), p(T)) = W(q_0, p_0) + \int_0^T \dot{W}(q(t), p(t))dt \qquad (10.7.6)$$

and choosing (q_0, p_0) in the sector where W is negative, but arbitrarily close to the origin.

Step 2 Employing an argument of Hahn [1967] to show spectral instability.

Here one uses the fact that ϵ is small and the original system is Hamiltonian. Indeed, the only nontrivial possibility to exclude for the eigenvalues on the imaginary axis is that they all stay there and are not zero for $\epsilon \neq 0$. Indeed, they cannot all move left by Step 1 and L_ϵ cannot have zero eigenvalues since $L_\epsilon z = 0$ implies $\dot{W}(z, z) = 0$. However, in this case, Hahn [1967] shows the existence of at least one periodic orbit, which cannot exist in view of (10.7.6) and the fact that $\dot{W}$ is negative definite.

This argument generalizes in two significant ways. First, it is valid in infinite dimensional systems, where M, S, R and Λ are replaced by linear operators. One of course needs some technical conditions to ensure that W has the requisite properties and that the evolution equations generate a semi-group on an appropriate Banach space. For Step 2 one requires, for example, that the spectrum at $\epsilon = 0$ be discrete with all eigenvalues having finite multiplicity.

The second generalization is to systems in block diagonal form but with a non-abelian group. The system (10.7.4) is the form that block diagonalization gives with an abelian symmetry group. For a non-abelian group, one gets, roughly speaking, a system consisting of (10.7.4) coupled with a Lie-Poisson (generalized rigid body) system. The main step needed in this case is a generalization of the Chetaev functional.

This formulation is attractive because of the interesting conclusions that can be obtained essentially from energetics alone. If one is willing to make *additional* assumptions, then there is a formula giving the amount by which *simple* eigenvalues move *off* the imaginary axis. One version of this formula, due to MacKay [1991] states that

$$\text{Re}\lambda_\epsilon = \frac{\bar{\xi}(\mathbb{J}B)_{\text{anti}}\xi}{\bar{\xi}^T \mathbb{J}\xi}\epsilon + O(\epsilon^2) \qquad (10.7.7)$$

where we write the linearized equations in the form

$$\dot{z} = L_\epsilon z = (\mathbb{J}Q + \epsilon B)z. \qquad (10.7.8)$$

λ_ϵ is the perturbed eigenvalue associated with a simple eigenvalue $\lambda_0 = i\omega_0$ on the imaginary axis at $\epsilon = 0$, ξ is a (complex) eigenvector for L_0 with eigenvalue λ_0, and $(\mathbb{J}B)_{\text{anti}}$ is the antisymmetric part of $\mathbb{J}B$.

In fact, the ratio of quadratic functions in (10.7.7) can be replaced by a ratio involving energy-like functions and their time derivatives including the energy itself or the Chetaev function. To actually work out (10.7.7) for examples like (10.7.1) can involve considerable calculation.

Here is a simple example in which one can carry out the entire analysis directly. We hasten to add that problems like the double spherical pendulum are considerably more complex algebraically and a direct analysis of the eigenvalue movement would not be so simple.

Consider the system (see Chetaev [1961])

$$\ddot{x} - g\dot{y} + \gamma\dot{x} + \alpha x = 0$$
$$\ddot{y} + g\dot{x} + \delta\dot{y} + \beta y = 0, \qquad (10.7.9)$$

which is a special case of (10.7.4). Assume $\gamma \geq 0$ and $\delta \geq 0$. For $\gamma = \delta = 0$ this system is Hamiltonian with symplectic form

$$\Omega = dx \wedge d\dot{x} + dy \wedge d\dot{y} - g\,dx \wedge dy \qquad (10.7.10)$$

and Hamiltonian

$$H = \frac{1}{2}(\dot{x}^2 + \dot{y}^2) + \frac{1}{2}(\alpha x^2 + \beta y^2). \qquad (10.7.11)$$

(Note that for $\alpha = \beta$, angular momentum is conserved.)

The characteristic polynomial is computed to be

$$p(\lambda) = \lambda^4 + (\gamma + \delta)\lambda^3 + (g^2 + \alpha + \beta + \gamma\delta)\lambda^2 + (\gamma\beta + \delta\alpha)\lambda + \alpha\beta. \qquad (10.7.12)$$

Let

$$p_0(\lambda) = \lambda^4 + (g^2 + \alpha + \beta)\lambda^2 + \alpha\beta. \qquad (10.7.13)$$

Since p_0 is a quadratic in λ^2, its roots are easily found. One gets:

i   If $\alpha > 0, \beta > 0$, then H is positive definite and the eigenvalues are on the imaginary axis; they are coincident in a 1:1 resonance for $\alpha = \beta$.

ii   If α and β have opposite signs, then H has index 1 and there is one eigenvalue pair on the real axis and one pair on the imaginary axis.

iii   If $\alpha < 0$ and $\beta < 0$ then H has index 2. Here the eigenvalues may or may not be on the imaginary axis.

To determine the cases, let

$$D = (g^2 + \alpha + \beta)^2 - 4\alpha\beta = g^4 + 2g^2(\alpha + \beta) + (\alpha - \beta)^2$$

be the discriminant. Then the roots of (10.7.13) are given by

$$\lambda^2 = \frac{1}{2}[-(g^2 + \alpha + \beta) \pm \sqrt{D}].$$

Thus we arrive at

a   If $D < 0$, then there are two roots in the right half plane and two in the left.

b   If $D = 0$ and $g^2 + \alpha + \beta > 0$, there are coincident roots on the imaginary axis, and if $g^2 + \alpha + \beta < 0$, there are coincident roots on the real axis.

c   If $D > 0$ and $g^2 + \alpha + \beta > 0$, the roots are on the imaginary axis and if $g^2 + \alpha + \beta < 0$, they are on the real axis.

Thus the case in which $D \geq 0$ and $g^2 + \alpha + \beta > 0$ (*i.e.*, if $g^2 + \alpha + \beta \geq 2\sqrt{\alpha\beta}$), is one to which the dissipation induced instabilities theorem applies.

Note that for $g^2 + \alpha + \beta > 0$, if D decreases through zero, a Hamiltonian Hopf bifurcation occurs. For example, as g increases and the eigenvalues move onto the imaginary axis, one speaks of the process as *gyroscopic stabilization*.

Now we add damping and get

Proposition 10.2 *If $\alpha < 0, \beta < 0, D > 0, g^2 + \alpha + \beta > 0$ and least one of γ, δ is strictly positive, then for (10.7.9), there is exactly one pair of eigenvalues in the strict right half plane.*

Proof We use the Routh-Hurwitz criterion (see Gantmacher [1959, vol. 2]), which states that the number of strict right half plane roots of the polynomial

$$\lambda^4 + \rho_1\lambda^3 + \rho_2\lambda^2 + \rho_3\lambda + \rho_4$$

equals the number of sign changes in the sequence

$$\left\{1, \rho_1, \frac{\rho_1\rho_2 - \rho_3}{\rho_1}, \frac{\rho_3(\rho_1\rho_2 - \rho_3) - \rho_1\rho_4}{\rho_1\rho_2 - \rho_3}, \rho_4\right\}. \qquad (10.7.14)$$

For our case, $\rho_1 = \gamma + \delta > 0, \rho_2 = g^2 + \alpha + \beta + \gamma\delta > 0, \rho_3 = \gamma\beta + \alpha\delta < 0$ and $\rho_4 = \alpha\beta > 0$, so the sign sequence (10.7.14) is

$$\{+, +, +, -, +\}.$$

Thus, there are two roots in the right half plane. ∎

It is interesting to speculate on the effect of damping on the Hamiltonian Hopf bifurcation in view of these general results and in particular, this example.

For instance, suppose $g^2 + \alpha + \beta > 0$ and we allow D to increase so a Hamiltonian Hopf bifurcation occurs in the undamped system. Then the above sign sequence does not change, so no bifurcation occurs in the damped system; the system is unstable and the Hamiltonian Hopf bifurcation just enhances it. However, if we simulate forcing or control by allowing one of γ or δ to be negative, but still small, then the sign sequence is more complex and one can get, for example, the Hamiltonian Hopf bifurcation breaking up into two nearly coincident Hopf bifurcations. This is consistent with the results of van Gils, Krupa and Langford [1990].

References

Abarbanel, H.D.I. and D.D. Holm [1987] Nonlinear stability analysis of inviscid flows in three dimensions: incompressible fluids and barotropic fluids, *Phys. Fluids* **30**, 3369–3382.

Abarbanel, H.D.I., D.D. Holm, J.E. Marsden and T.S. Ratiu [1986] Nonlinear stability analysis of stratified fluid equilibria, *Phil. Trans. R. Soc. Lond. A* **318**, 349–409; also *Phys. Rev. Lett.* **52** [1984] 2352–2355.

Abed, E.H. and J-H. Fu [1986] Local feedback stabilization and bifurcation control. *Syst. and Cont. Lett.* **7**, 11-17, **8**, 467-473.

Abraham, R. and J.E. Marsden [1978] *Foundations of Mechanics.* Second Edition, Addison-Wesley Publishing Co., Reading, Mass..

Abraham, R., J.E. Marsden and T.S. Ratiu [1988] *Manifolds, Tensor Analysis, and Applications.* Second Edition, Springer-Verlag, New York.

Adams, M.R., J. Harnad and E. Previato [1988] Isospectral Hamiltonian flows in finite and infinite dimensions I. Generalized Moser systems and moment maps into loop algebras, *Comm. Math. Phys.* **117**, 451–500.

Aeyels, D. and M. Szafranski [1988] Comments on the stabilizability of the angular velocity of a rigid body, *Syst. and Cont. Lett.* **10**, 35–39.

Aharonov, Y. and J. Anandan [1987] Phase change during acyclic quantum evolution, *Phys. Rev. Lett.* **58**, 1593–1596.

Alber, M. and J.E. Marsden [1991] On geometric phases for soliton equations, *preprint*.

Anandan, J. [1988] Geometric angles in quantum and classical physics, *Phys. Lett.* **A 129**, 201–207.

Andrews, D.G. [1984] On the existence of nonzonal flows satisfying sufficient conditions for stability, *Geo. Astr. Fluid Dyn.* **28**, 243–256.

Armbruster, D., J. Guckenheimer and P. Holmes [1988] Heteroclinic cycles and modulated traveling waves in systems with O(2)-symmetry, *Physica D* **29**, 257–282.

Arms, J.M. [1981] The structure of the solution set for the Yang-Mills equations, *Math. Proc. Comb. Phil. Soc.* **90**, 361–372.

Arms, J.M., A. Fischer and J.E. Marsden [1975] Une approche symplectique pour des théorémes de décomposition en géométrie ou relativité générale, *C. R. Acad. Sci. Paris* **281**, 517–520.

Arms, J.M., J.E. Marsden and V. Moncrief [1981] Symmetry and bifurcations of momentum mappings, *Comm. Math. Phys.* **78**, 455–478.

Arms, J.M., J.E. Marsden and V. Moncrief [1982] The structure of the space solutions of Einstein's equations: II Several Killing fields and the Einstein-Yang-Mills equations, *Ann. of Phys.* **144**, 81–106.

Arnold, V.I. [1964] Instability of dynamical systems with several degrees of freedom, *Dokl. Akad. Nauk. SSSR* **156**, 9–12.

Arnold, V.I. [1966] Sur la géometrie differentielle des groupes de Lie de dimenson infinie et ses applications à l'hydrodynamique des fluids parfaits, *Ann. Inst. Fourier, Grenoble* **16**, 319–361.

Arnold, V.I. [1969] On an a priori estimate in the theory of hydrodynamical stability, *Am. Math. Soc. Transl.* **79**, 267–269.

Arnold, V.I. [1988] *Dynamical Systems III.* Encyclopedia of Mathemaics **3**, Springer-Verlag.

Arnold, V.I. [1989] *Mathematical Methods of Classical Mechanics.* Second Edition, Graduate Texts in Math. **60**, Springer-Verlag.

Ashbaugh, M.S., C.C. Chicone and R.H. Cushman [1991] The twisting tennis racket, *Dyn. and Diff. Eqn's.* **3**, 67–85.

Atiyah, M. [1982] Convexity and commuting Hamiltonïans, *Bull. Lon. Math. Soc.* **14**, 1–15.

Austin, M., P.S. Krishnaprasad and L.S. Wang [1991] Symplectic and almost Poisson integration, *preprint*.

Baider, A., R.C. Churchill and D.L. Rod [1990] Monodromy and nonintegrability in complex Hamiltonian systems, *J. Dynamics and Differential Equations* **2**, 451–481.

Baillieul, J. [1987] Equilibrium mechanics of rotating systems, *Proc. CDC* **26**, 1429–1434.

Baillieul, J. and M. Levi [1987] Rotational elastic dynamics, *Physica D* **27**, 43–62.

Baillieul, J. and M. Levi [1991] Constrained relative motions in rotational mechanics, *Arch. Rat. Mech. An.* **115**, 101–135.

Ball, J.M. and J.E. Marsden [1984] Quasiconvexity at the boundary, positivity of the second variation and elastic stability, *Arch. Rat. Mech. An.* **86**, 251–277.

Bao, D., J.E. Marsden and R. Walton [1984] The Hamiltonian structure of general relativistic perfect fluids, *Comm. Math. Phys.* **99**, 319–345.

Batt, J. and G. Rein [1991] A rigorous stability result for the Vlasov-Poisson system in three dimensions, *Ann. Math. Pura Appl.* (to appear).

Benjamin, T.B. [1984] Impulse, flow force and variational principles, *IMA J. of Appl. Math.* **32**, 3–68.

Benjamin, T.B. and P.J. Olver [1982] Hamiltonian structure, symmetries and conservation laws for water waves, *J. Fluid Mech.* **125**, 137–185.

Berry, M. [1984] Quantal phase factors accompanying adiabatic changes, *Proc. Roy. Soc. London A* **392**, 45–57.

Berry, M. [1985] Classical adiabatic angles and quantal adiabatic phase, *J. Phys. A. Math. Gen.* **18**, 15–27.

Berry, M. and J. Hannay [1988] Classical non-adiabatic angles, *J. Phys. A. Math. Gen.* **21**, 325–333.

Bertrand, J. [1852] *J. de Math.* **27**, 393.

Bloch, A.M.[1989] Steepest descent, linear programming and Hamiltonian flows, *Cont. Math. AMS* **114**, 77–88.

Bloch, A.M., R.W. Brockett and T.S. Ratiu [1990] A new formulation of the generalized Toda Lattice equations and their fixed point analysis via the momentum map, *Bull. Amer. Math. Soc.* **23**, 477–485.

Bloch, A.M., P.S. Krishnaprasad, J.E. Marsden and T.S. Ratiu [1991] Dissipation induced instabilities, (to appear).

Bloch, A.M., P.S. Krishnaprasad, J.E. Marsden and G. Sánchez de Alvarez [1991] Stabilization of rigid body dynamics by internal and external torques, *Automatica* (to appear).

Bloch, A.M. and J.E. Marsden [1989] Controlling homoclinic orbits, *Theor. and Comp. Fluid Mech.* **1**, 179–190.

Bloch, A.M. and J.E. Marsden [1990] Stabilization of rigid body dynamics by the energy-Casimir method, *Syst. and Cont. Lett.* **14**, 341–346.

Bloch, A.M., N.H. McClamroch and M. Reyhanoglu [1990] Controllability and stabilizability properties of a nonholonomic control system, *Proc. CDC, Hawaii*, 1312–1314.

Bobenko, A.I., A.G. Reyman and M.A. Semenov-Tian-Shansky [1989] The Kowalewski Top 99 years later: A Lax pair, generalizations and explicit solutions, *Comm. Math. Phys.* **122**, 321–354.

Bonnard, B. [1986] Controllabilité des systèmes méchanique sur les groups de Lie , *SIAM J. Cont. and Optim.* **22**, 711–722.

Born, M. and L. Infeld [1935] On the quantization of the new field theory, *Proc. Roy. Soc. A* **150**, 141.

Brockett, R.W. [1973] Lie algebras and Lie groups in control theory, in *Geometric Methods in Systems Theory*, Proc. NATO Advanced Study Institute, R.W. Brockett and D.Q. Mayne (eds.), Reidel, 43–82.

Brockett, R.W. [1976] Nonlinear systems and differential geometry, *Proc. IEEE* **64**, No 1, 61–72.

Brockett, R.W. [1981] Control theory and singular Riemannian geometry, in *New Directions in Applied Mathematics*, P.J. Hilton and G.S. Young (eds.), Springer.

Brockett, R.W. [1983] Asymptotic stability and feedback stabilization, in *Differential Geometric Control Theory*, R.W. Brockett, R.S. Millman and H. Sussman (eds.), Birkhauser.

Brockett, R.W. [1987] On the control of vibratory actuators, *Proc. 1987 IEEE Conf. Decision and Control*, 1418–1422.

Brockett, R.W. [1989a] On the rectification of vibratory motion, *Sensors and Actuators* **20**, 91–96.

Brockett, R.W. [1989b] Dynamical systems that sort lists and solve linear programming problems, *Lin. Alg. and Appl.* **146**, 79–91.

Buttke, T. [1991] (preprint).

Byrnes, C.I. and A. Isidori [1989] Attitude stabilization of rigid spacecraft, *Automatica*, to appear.

Carnevale, G.F. and T.G. Shepherd [1990] On the interpretation of Andrew's theorem, *Geo. Astro. Fluid. Dyn.* **51**, 4–7.

Cartan, E. [1923] Sur les varietes a connexion affine et theorie de relativité généralizée, *Ann. Ecole Norm. Sup.* **40**, 325–412, **41**, 1–25.

Cartan, E. [1928] Sur la stabilité ordinaire des ellipsoides de Jacobi, *Proc. Int. Math. Cong. Toronto* **2**, 9–17.

Cendra, H. and J.E. Marsden [1987] Lin constraints, Clebsch potentials and variational principles, *Physica D* **27**, 63–89.

Cendra, H., A. Ibort and J.E. Marsden [1987] Variational principal fiber bundles: a geometric theory of Clebsch potentials and Lin constraints, *J. of Geom. and Phys.* **4**, 183–206.

Chandrasekhar, S. [1977] *Ellipsoidal Figures of Equilibrium.* Dover.

Chandrasekhar, S. [1961] *Hydrodynamic and Hydromagnetic Instabilities.* Oxford University Press.

Channell, P. and C. Scovel [1990] Symplectic integration of Hamiltonian systems, *Nonlinearity* **3**, 231–259.

Chern, S.J. [1991] *Mathematical Theory of the Barotropic Model in Geophysical Fluid Dynamics.* Thesis, Cornell University.

Chern, S.J. and J.E. Marsden [1990] A note on symmetry and stability for fluid flows, *Geo. Astro. Fluid. Dyn.* **51**, 1–4.

Chernoff, P.R. and J.E. Marsden [1974] *Properties of infinite dimensional Hamiltonian systems.* Lecture Notes in Math. **425**, Springer-Verlag, New York.

Chetayev, N.G. [1961] *The Stability of Motion.* Trans. by M. Nadler Pergamon Press, New York.

Chillingworth, D.R.J., J.E. Marsden and Y.H. Wan [1982] Symmetry and bifurcation in three dimensional elasticity, *Arch. Rat. Mech. An.* **80**, 295–331, **83** [1983] 363–395, **84** [1984] 203–233.

Chirikov, B.V. [1979] A universal instability of many dimensional oscillator systems, *Phys. Rep.* **52**, 263–379.

Chorin, A.J, T.J.R. Hughes, J.E. Marsden and M. McCracken [1978] Product Formulas and Numerical Algorithms, *Comm. Pure and Appl. Math.* **31**, 205–256.

Chorin, A.J. and J.E. Marsden [1989] *A Mathematical Introduction to Fluid Mechanics.* Texts in Applied Mathematics, **4**, Springer-Verlag, New York.

Chow, S.N. and J.K. Hale [1982] *Methods of Bifurcation Theory.* Springer, New York.

Clebsch, A. [1857] Über eine allgemeine Transformation der hydrodynamischen Gleichungen, *Zeitschrift für Reine Angew. Math.* **54**, 293–312.

Clebsch, A. [1859] Über die Integration der hydrodynamischen Gleichungen, *Zeitschrift für Reine Angew. Math.* **56**, 1–10.

Cordani, B. [1986] Kepler problem with a magnetic monopole, *J. Math. Phys.* **27**, 2920–2921.

Crouch, P.E. [1986] Spacecraft attitude control and stabilization: Application of geometric control to rigid body models, *IEEE Trans. Aut. Cont.* **AC-29**, 321–331.

Cushman, R. and D. Rod [1982] Reduction of the semi-simple 1:1 resonance, *Physica D* **6**, 105–112.

Darling, B.T. and D.M. Dennison [1940] The water vapor molecule, *Phys. Rev.* **57**, 128–139.

David, D. and D.D. Holm [1990] Multiple Lie-Poisson structures, reductions, and geometric phases for the Maxwell-Bloch travelling-wave equations, *preprint.*

David, D., D.D. Holm and M. Tratnik [1990] Hamiltonian chaos in nonlinear optical polarization dynamics, *Physics reports* **187**, 281–370.

Dellnitz, M., J.E. Marsden, I. Melbourne and J. Scheurle [1992] (in preparation).

Dellnitz, M., I. Melbourne and J.E. Marsden [1992] Generic bifurcation of Hamiltonian vector fields with symmetry, (to appear).

Deprit, A. [1983] Elimination of the nodes in problems of N bodies, *Celestial Mech.* **30**, 181–195.

deVogelaére, R. [1956] Methods of integration which preserve the contact transformation property of the Hamiltonian equations, *Department of Mathematics, University of Notre Dame Report* **4**.

Dirac, P.A.M. [1964] Lectures on quantum mechanics, *Belfer Graduate School of Science, Monograph Series* **2**, Yeshiva University.

Dritchel, D. [1989] The stability of elliptical vortices in an external straining flow, *J. Fluid Mech.* **210**, 223–261.

Duistermaat, J.J. [1980] On global action angle coordinates, *Comm. Pure Appl. Math.* **33**, 687–706.

Dzyaloshinskii, I.E. and G.E. Volovick [1980] Poisson brackets in condensed matter physics, *Ann. of Phys.* **125**, 67–97.

Ebin, D.G. and J.E. Marsden [1970] Groups of diffeomorphisms and the motion of an incompressible fluid, *Ann. Math.* **92**, 102–163.

Eckart, C. [1935] Some studies concerning rotating axes and polyatomic molecules, *Phys. Rev.* **47**, 552–558.

Emmrich, C. and H. Römer [1990] Orbifolds as configuration spaces of systems with gauge symmetries, *Comm. Math. Phys.* **129**, 69–94.

Ercolani, N., M.G. Forest and D.W. McLaughlin [1990] Geometry of the modulational instability, III. Homoclinic orbits for the periodic sine-Gordon equation, *Physica D* **43**, 349.

Faybusovich, L. [1991] Hamiltonian structure of dynamical systems which solve linear programing problems, *preprint*.

Feng, K. [1986] Difference Schemes for Hamiltonian Formalism and Symplectic Geometry, *J. Comp. Math.* **4**, 279–289.

Feng, K. and Z. Ge [1988] On approximations of Hamiltonian systems, *J. Comp. Math.* **6**, 88–97.

Feng, K. and M.Z. Qin [1987] The symplectic methods for the computation of Hamiltonian equations, *Springer Lecture Notes in Math.* **1297**, 1–37.

Fischer, A.E. and J.E. Marsden [1972] The Einstein equations of evolution — a geometric approach, *J. Math. Phys.* **13**, 546–68.

Fischer, A.E. and J.E. Marsden [1979] Topics in the dynamics of general relativity, in *Isolated Gravitating systems in General Relativity*, J. Ehlers (ed.), Italian Physical Society, 322–395.

Fischer, A.E., J.E. Marsden and V. Moncrief [1980] The structure of the space of solutions of Einstein's equations, I: One Killing field, *Ann. Inst. H. Poincaré* **33**, 147–194.

Fontich, E. and J.C. Simo [1990] The splitting of separatrices for analytic diffeomorphisms, *Ergodic. Th. and Dyn. Sys.* **10**, 295–318.

Galin, D.M. [1982] Versal deformations of linear Hamiltonian systems, *AMS Transl.* **118**, 1–12, (1975 Trudy Sem. Petrovsk. **1**, 63–74).

Ge, Z. [1990] Generating functions, Hamilton-Jacobi equation and symplectic groupoids over Poisson manifolds, *Ind. Univ. Math. J.* **39**, 859–876.

Ge, Z. [1991a] Equivariant symplectic difference schemes and generating functions, *Physica D* **49**, 376–386.

Ge, Z. [1991b] A constrained variational problem and the space of horizontal paths, *Pac. J. Math.* **149**, 61–94.

Ge, Z. and J.E. Marsden [1988] Lie-Poisson integrators and Lie-Poisson Ham-iltonian-Jacobi theory, *Phys. Lett. A* **133**, 134–139.

Goldman, W.M. and J.J. Millson [1990] Differential graded Lie algebras and singularities of level sets of momentum mappings, *Comm. Math. Phys.* **131**, 495–515.

Golin, S., A. Knauf and S. Marmi [1989] The Hannay angles: geometry, adiabaticity, and an example, *Comm. Math. Phys.* **123**, 95–122.

Golin, S. and S. Marmi [1990] A class of systems with measurable Hannay angles, *Nonlinearity* **3**, 507–518.

Golubitsky, M. and I. Stewart [1987] Generic bifurcation of Hamiltonian systems with symmetry, *Physica D* **24**, 391–405.

Golubitsky, M., I. Stewart and D. Schaeffer [1988] *Singularities and Groups in Bifurcation Theory, Vol. 2.* Springer-Verlag, AMS Series **69**.

Gotay, M., J. Isenberg, J.E. Marsden and R. Montgomery [1992] *Momentum Maps and the Hamiltonian Structure of Classical Relativistic Field Theories.* Springer-Verlag, (to appear).

Gotay, M., R. Lashof, J. Sniatycki and A. Weinstein [1980] Closed forms on symplectic fiber bundles, *Comm. Math. Helv.* **58**, 617–621.

Greenspan, D. [1974] *Discrete Numerical Methods in Physics and Engineering.* Academic Press.

Greenspan, D. [1984] Conservative numerical methods for $\ddot{x} = f(x)$, *J. Comp. Phys.* **56**, 21–48.

Grossman, R., P.S. Krishnaprasad and J.E. Marsden [1988] The dynamics of two coupled rigid bodies, in *Dynamical Systems Approaches to Nonlinear Problems in Systems and Circuits*, Salam and Levi (eds.), *SIAM*, 373–378.

Guckenheimer, J. and P. Holmes [1983] *Nonlinear Oscillations, Dynamical Systems and Bifurcations of vector Fields.* Springer Applied Math. Sciences **43**.

Guckenheimer, J. and P. Holmes [1988] Structurally stable heteroclinic cycles, *Math. Proc. Cam. Phil. Soc.* **103**, 189–192.

Guichardet, A. [1984] On rotation and vibration motions of molecules, *Ann. Inst. H. Poincaré* **40**, 329–342.

Guillemin, V. and E. Prato [1990] Heckman, Kostant, and Steinberg formulas for symplectic manifolds, *Adv. in Math.* **82**, 160–179.

Guillemin, V. and S. Sternberg [1980] The moment map and collective motion, *Ann. of Phys.* **1278**, 220–253.

Guillemin, V. and S. Sternberg [1984] *Symplectic Techniques in Physics.* Cambridge University Press.

Hahn, W. [1967] *Stability of Motion.* Springer-Verlag, New York.

Hannay, J. [1985] Angle variable holonomy in adiabatic excursion of an itegrable Hamiltonian, *J. Phys. A: Math. Gen.* **18**, 221–230.

Harnad, J. and J. Hurtubise [1990] Generalized tops and moment maps into loop algebras, *preprint, CRM*.

Harnad, J., J. Hurtubise and J.E. Marsden [1991] Reduction of Hamiltonian systems with discrete symmetry, *preprint*.

Hirschfelder, J.O. and J.S. Dahler [1956] The kinetic energy of relative motion, *Proc. Nat. Acad. Sci.* **42**, 363–365.

Hodgins, J. and M. Raibert [1990] Biped gymnastics, *Int. J. Robotics Research*, (in press).

Holm, D.D., B.A. Kupershmidt and C.D. Levermore [1985] Hamiltonian differencing of fluid dynamics, *Adv. in Appl. Math.* **6**, 52–84.

Holm, D.D. and J.E. Marsden [1991] The rotor and the pendulum, *Proc. Souriaufest*, Birkhauser.

Holm, D.D., J.E. Marsden and T.S. Ratiu [1986] The Hamiltonian structure of continuum mechanics in material, spatial and convective representations, *Séminaire de Mathématiques supérieurs, Les Presses de L'Univ. de Montréal* **100**, 11–122.

Holm, D.D., J.E. Marsden, T.S. Ratiu and A. Weinstein [1985] Nonlinear stability of fluid and plasma equilibria, *Phys. Rep.* **123**, 1–116.

Holmes, P.J. and J.E. Marsden [1981] A partial differential equation with infinitely many periodic orbits: chaotic oscillations of a forced beam, *Arch. Rat. Mech. Anal.* **76**, 135–166.

Holmes, P.J. and J.E. Marsden [1982a] Horseshoes in perturbations of Hamiltonian systems with two degrees of freedom, *Comm. Math. Phys.* **82**, 523–544.

Holmes, P.J. and J.E. Marsden [1982b] Melnikov's method and Arnold diffusion for perturbations of integrable Hamiltonian systems, *J. Math. Phys.* **23**, 669–675.

Holmes, P.J. and J.E. Marsden [1983] Horseshoes and Arnold diffusion for Hamiltonian systems on Lie groups, *Indiana Univ. Math. J.* **32**, 273–310.

Holmes, P.J., J.E. Marsden and J. Scheurle [1988] Exponentially small splittings of separatrices with applications to KAM theory and degenerate bifurcations, *Cont. Math.* **81**, 213–244.

Howard, J.E. and R.S. Mackay [1987] Linear stability of symplectic maps, *J. Math. Phys.* **28**, 1036–1051.

Howard, J.E. and R.S. Mackay [1987] Calculation of linear stability boundaries for equilibria of Hamiltonian systems, *Phys. Lett. A.* **122**, 331–334.

Iwai, T. [1982] On a "conformal" Kepler problem and its reduction, *J. Math. Phys.* **22**, 1633–1639.

Iwai, T. [1982] The symmetry group of the harmonic oscillator and its reduction, *J. Math. Phys.* **23**, 1088–1092.

Iwai, T. [1987a] A gauge theory for the quantum planar three-body system, *J. Math. Phys.* **28**, 1315–1326.

Iwai, T. [1987b] A geometric setting for internal motions of the quantum three-body system, *J. Math. Phys.* **28**, 1315–1326.

Iwai, T. [1987c] A geometric setting for classical molecular dynamics, *Ann. Inst. Henri Poincaré, Phys. Th.* **47**, 199–219.

Iwai, T. [1990a] On the Guichardet/Berry connection, *Phys. Lett. A* **149**, 341–344.

Iwai, T. [1990b] The geometry of the $SU(2)$ Kepler problem, *J. Geom. and Phys.* **7**, 507–535.

Iwai, T. and Y. Uwano [1991] The MIC-Kepler problem and its symmetry group for positive energies both in classical and quantum mechanics, *Il Nuovo Cimento*, **106**, 849–878.

Iwiński, Z.R. and L.A. Turski [1976] Canonical theories of systems interacting electromagnetically, *Letters in Applied and Engineering Sciences* **4**, 179–191.

Jacobi, C.G.K. [1843] *J. für Math.* **26**, 115.

Jellinek, J. and D.H. Li [1989] Separation of the energy of overall rotations in an N-body system, *Phys. Rev. Lett.* **62**, 241–244.

Jepson, D.W. and J.O. Hirschfelder [1958] Set of coordinate systems which diagonalize the kinetic energy of relative motion, *Proc. Nat. Acad. Sci.* **45**, 249–256.

Kane, T.R. and M. Scher [1969] A dynamical explanation of the falling cat phenomenon, *Int. J. Solids Structures* **5**, 663–670.

Kazhdan, D., B. Kostant and S. Sternberg [1978] Hamiltonian group actions and dynamical systems of Calogero type, *Comm. Pure Appl. Math.* **31**, 481–508.

Kirillov, A.A. [1962] Unitary representations of nilpotent Lie groups, *Russian Math. Surveys* **17**, 53–104.

Kirillov, A.A. [1976] *Elements of the Theory of Representations.* Springer-Verlag, New York.

Klein, F. [1897] *The Mathematical Theory of the Top.* Scribner's, New York.

Koiller, J. and M.J.H. Dantes [1990] Motion of impurities in ideal planar fluids, *preprint.*

Kopell, N. and R.B. Washburn, Jr. [1982] Chaotic motions in the two degree-of-freedom swing equations, *IEEE Trans. Circuits Syst.*, 738–746.

Kosmann-Schwarzbach, Y. and F. Magri [1990] Poisson-Nijenhuis structures, *Ann. Inst. H. Poincaré* **53**, 35–81.

Krein, M.G. [1950] A generalization of several investigations of A.M. Liapunov on linear differential equations with periodic coefficients, *Dokl. Akad. Nauk. SSSR* **73**, 445–448.

Krishnaprasad, P.S. [1985] Lie-Poisson structures, dual-spin spacecraft and asymptotic stability, *Nonl. An. Th. Meth. and Appl.* **9**, 1011–1035.

Krishnaprasad, P.S. [1989] Eulerian many-body problems, *Cont. Math. AMS* **97**, 187–208.

Krishnaprasad, P.S. and J.E. Marsden [1987] Hamiltonian structure and stability for rigid bodies with flexible attachments, *Arch. Rat. Mech. An.* **98**, 137–158.

Krishnaprasad, P.S. and L.S. Wang [1992] Gyroscopic control and stabilization, *J. Nonlinear Sci.* (to appear).

Krishnaprasad, P.S. and R. Yang [1991] Geometric phases, anholonomy and optimal movement, *Proc. IEEE Conf. on Rob. and Aut.*, 1–5.

Krishnaprasad, P.S., R. Yang, and W.P. Dayawansa [1991] Control problems on principle bundles and nonholonomic mechanics, *preprint.*

Krupa, M. [1990] Bifurcations of relative equilibria, *SIAM J. Math. An.* **21**, 1453–1486.

Kruskal, M. and H. Segur [1991] Asymptotics beyond all orgers in a model of crystal growth. *Stud. in Appl. Math.* **85**, 129–181.

Kummer, M. [1981] On the construction of the reduced phase space of a Hamiltonian system with symmetry, *Indiana Univ. Math. J.* **30**, 281–291.

Levi, M. [1989] Morse theory for a model space structure, *Cont. Math. AMS* **97**, 209–216

Lewis, D.R. [1989] Nonlinear stability of a rotating planar liquid drop, *Arch. Rat. Mech. Anal.* **106**, 287–333.

Lewis, D.R. [1991] Lagrangian block diagonalization, *Dyn. Diff. Eqn's.* (to appear).

Lewis, D.R., J.E. Marsden, R. Montgomery and T.S. Ratiu [1986] The Hamiltonian structure for dynamic free boundary problems, *Physica D* **18**, 391–404.

Lewis, D.R., J.E. Marsden and T.S. Ratiu [1987] Stability and bifurcation of a rotating liquid drop, *J. Math. Phys.* **28**, 2508–2515.

Lewis, D.R., J.E. Marsden, T.S. Ratiu and J.C. Simo [1990] Normalizing connections and the energy-momentum method, Proceedings of the CRM conference on *Hamiltonian systems, Transformation Groups, and Spectral Transform Methods*, CRM Press, Harnad and Marsden (eds.), 207–227.

Lewis, D.R., T.S. Ratiu, J.C. Simo and J.E. Marsden [1992] The heavy top, a geometric treatment, *Nonlinearity* (to appear).

Lewis, D.R. and J.C. Simo [1990] Nonlinear stability of rotating pseudo-rigid bodies, *Proc. Roy. Soc. Lon. A* **427**, 281–319.

Li, C.W. and Qin, M.Z. [1988] A symplectic difference scheme for the infinite dimensional Hamilton system, *J. Comp. Math.* **6**, 164–174.

Li, Z., L. Xiao and M.E. Kellman [1990] Phase space bifurcation structure and the generalized local-to-normal transition in resonantly coupled vibrations, *J. Chem. Phys.* **92**, 2251–2268.

Lichtenberg, A.J. and M.A. Liebermann [1983] Regular and stochastic motion, *Applied Math. Science* **38**, Springer-Verlag.

Lie, S. [1890] *Theorie der Transformationsgruppen, Zweiter Abschnitt.* Teubner, Leipzig.

Littlejohn, R.G. [1988] Cyclic evolution in quantum mechanics and the phases of Bohr-Sommerfeld and Maslov, *Phys. Rev. Lett.* **61**, 2159–2162.

Low, F.E. [1958] A Lagrangian formulation of the Boltzmann-Vlasov equation for plasmas, *Proc. Roy. Soc. Lond. A* **248**, 282–287.

Lu, J.H. and T.S. Ratiu [1991] On Kostant's convexity theorem, *J. of the AMS* (to appear).

MacKay, R. [1986] Stability of equilibria of Hamiltonian systems, in *Nonlinear Phenomena and Chaos.* S. Sarkar (ed.), 254–270.

MacKay, R. and P. Saffman [1986] Stability of water waves, *Proc. Roy. Soc. Lon. A* **406**, 115–125.

MacKay, R. [1987] Instability of vortex streets, *Dyn. Stab. Syst.* **2**, 55–71.

MacKay, R. [1991a] Movement of eigenvalues of Hamiltonian equilibria under non-Hamiltonian perturbation, *Phys. Lett. A* **155**, 266–268.

MacKay, R. [1991b] Some aspects of the dynamics and numerics of Hamiltonian systems, *preprint.*

Marle, C.M. [1976] Symplectic manifolds, dynamical groups and Hamiltonian mechanics, in *Differential Geometry and Relativity*, M. Cahen and M. Flato (eds.), D. Reidel, Boston.

Marsden, J.E. [1981] *Lectures on Geometric Methods in Mathematical Physics.* SIAM, Philadelphia, PA.

Marsden, J.E. [1982] A group theoretic approach to the equations of plasma physics, *Can. Math. Bull.* **25**, 129–142.

Marsden, J.E. [1987] *Appendix to* Golubitsky and Stewart [1987].

Marsden, J.E. and T.J.R. Hughes [1983] *Mathematical Foundations of Elasticity.* Prentice-Hall, Redwood City, Calif.

Marsden, J.E., R. Montgomery, P. Morrison and W.B. Thompson [1986] Covariant Poisson brackets for classical fields, *Ann. of Phys.* **169**, 29–48.

Marsden, J.E., R. Montgomery and T.S. Ratiu [1990] *Reduction, symmetry, and phases in mechanics.* Memoirs AMS **436**.

Marsden, J.E., P.J. Morrison and A. Weinstein [1984] The Hamiltonian structure of the BBGKY hierarchy equations, *Cont. Math. AMS* **28**, 115–124.

Marsden, J.E., O.M. O'Reilly, F.J. Wicklin and B.W. Zombro [1991] Symmetry, stability, geometric phases, and mechanical integrators, *Nonlinear Science Today* **1**, 4–11, **1**, 14–21.

Marsden, J.E. and T.S. Ratiu [1986] Reduction of Poisson manifolds, *Lett. in Math. Phys.* **11**, 161–170.

Marsden, J.E. and T.S. Ratiu [1992] *Symmetry and Mechanics.* Springer-Verlag (in preparation).

Marsden, J.E., T.S. Ratiu and G. Raugel [1991] Symplectic connections and the linearization of Hamiltonian systems, *Proc. Roy. Soc. Ed. A* **117**, 329-380

Marsden, J.E., T.S. Ratiu and A. Weinstein [1984a] Semi-direct products and reduction in mechanics, *Trans. Am. Math. Soc.* **281**, 147–177.

Marsden, J.E., T.S. Ratiu and A. Weinstein [1984b] Reduction and Hamiltonian structures on duals of semidirect product Lie Algebras, *Cont. Math. AMS* **28**, 55–100.

Marsden, J.E. and J. Scheurle [1992] Lagrangian reduction and the double spherical pendulum, *preprint.*

Marsden, J.E., J.C. Simo, D.R. Lewis and T.A. Posbergh [1989] A block diagonalization theorem in the energy momentum method, *Cont. Math. AMS* **97**, 297–313.

Marsden, J.E. and A. Weinstein [1974] Reduction of symplectic manifolds with symmetry, *Rep. Math. Phys.* **5**, 121–130.

Marsden, J.E. and A. Weinstein [1982] The Hamiltonian structure of the Maxwell-Vlasov equations, *Physica D* **4**, 394–406.

Marsden, J.E. and A. Weinstein [1983] Coadjoint orbits, vortices and Clebsch variables for incompressible fluids, *Physica D* **7**, 305–323.

Marsden, J.E., A. Weinstein, T.S. Ratiu, R. Schmid and R.G. Spencer [1983] Hamiltonian systems with symmetry, coadjoint orbits and plasma physics, in Proc. IUTAM-IS1MM Symposium on *Modern Developments in Analytical Mechanics*, Torino 1982, *Atti della Acad. della Sc. di Torino* **117**, 289–340.

Martin, J.L. [1959] Generalized classical dynamics and the "classical analogue" of a Fermi oscillation, *Proc. Roy. Soc. A* **251**, 536.

Melnikov, V.K. [1963] On the stability of the center for time periodic perturbations, *Trans. Moscow Math. Soc.* **12**, 1–57.

Meyer, K.R. [1973] Symmetries and integrals in mechanics, in *Dynamical Systems*, M. Peixoto (ed.), Academic Press, 259–273.

Meyer, K.R. [1981] Hamiltonian systems with a discrete symmetry, *J. Diff. Eqn's.* **41**, 228–238.

Meyer, K.R. and R. Hall [1991] *Hamiltonian Mechanics and the n-body Problem.* Springer-Verlag, Applied Mathematical Sciences.

Misner, C., K. Thorne and J.A. Wheeler [1973] *Gravitation.* W.H. Freeman, San Francisco.

Montaldi, J.A., R.M. Roberts and I.N. Stewart [1988] Periodic solutions near equilibria of symmetric Hamiltonian systems, *Phil. Trans. R. Soc. Lond. A* **325**, 237–293.

Montgomery, R. [1984] Canonical formulations of a particle in a Yang-Mills field, *Lett. Math. Phys.* **8**, 59–67.

Montgomery, R [1986] *The bundle picture in mechanics* Thesis, UC Berkeley.

Montgomery, R. [1988] The connection whose holonomy is the classical adiabatic angles of Hannay and Berry and its generalization to the non-integrable case, *Comm. Math. Phys.* **120**, 269–294.

Montgomery, R. [1990] Isoholonomic problems and some applications, *Comm. Math Phys.* **128**, 565–592.

Montgomery, R. [1991] How much does a rigid body rotate? A Berry's phase from the 18th century, *Am. J. Phys.* **59**, 394–398.

Montgomery, R., J.E. Marsden and T.S. Ratiu [1984] Gauged Lie-Poisson structures, *Cont. Math. AMS* **28**, 101–114.

Montgomery, R., M. Raibert and Zexing Li [1990] Dynamics and optimal control of legged locomotion systems, *preprint*.

Moon, F.C. and P.J. Holmes [1979] A magnetoelastic strange attractor, *J. of Sound and Vib.* **65**, 275–296, **69**, 339.

Morozov, V.M., V.N. Rubanovskii, V.V. Rumiantsev and V.A. Samsonov [1973] On the bifurcation and stability of the steady state motions of complex mechanical systems, *PMM* **37**, 387–399.

Morrison, P.J. [1980] The Maxwell-Vlasov equations as a continuous Hamiltonian system, *Phys. Lett. A* **80**, 383–386.

Morrison, P.J. [1982] Poisson Brackets for fluids and plasmas, in *Mathematical Methods in Hydrodynamics and Integrability in Related Dynamical Systems*, AIP Conf. Proc. **88**, M. Tabor and Y.M. Treve (eds.), La Jolla, Calif.

Morrison, P.J. [1986] A paradigm for joined Hamiltonian and dissipative systems, *Physica D* **18**, 410–419.

Morrison, P.J. [1987] Variational principle and stability of nonmonotone Vlasov-Poisson equilibria, *Z. Naturforsch.* **42a**, 1115–1123.

Morrison, P.J. and S. Eliezer [1986] Spontaneous symmetry breaking and neutral stability on the noncanonical Hamiltonian formalism, *Phys. Rev. A* **33**, 4205.

Morrison, P.J. and J.M. Greene [1980] Noncanonical Hamiltonian density formulation of hydrodynamics and ideal magnetohydrodynamics, *Phys. Rev. Lett.* **45**, 790–794, errata **48** (1982), 569.

Morrison, P.J. and R.D. Hazeltine [1984] Hamiltonian formulation of reduced magnetohydrodynamics, *Phys. Fluids* **27**, 886–897.

Morrison, P.J. and D. Pfirsch [1990] The free energy of Maxwell-Vlasov equilibria, *Phys. Fluids B* **2**, 1105–1113.

Moser, J. [1958] New aspects in the theory of stability of Hamiltonian systems, *Comm. on Pure and Appl. Math.* **XI**, 81–114.

Moser, J. [1965] On the volume elements on a manifold, *Trans. Am. Soc.* **120**, 286–294.

Moser, J. [1976] Periodic orbits near equilibrium and a theorem by Alan Weinstein, *Comm. Pure Appl. Math.* **29**, 727–747.

Nambu, Y. [1973] Generalized Hamiltonian dynamics, *Phys. Rev. D* **7**, 2405–2412.

Neishtadt, A. [1984] The separation of motions in systems with rapidly rotating phase, *P.M.M. USSR* **48**, 133–139.

Noether, E. [1918] Invariante Variationsprobleme, *Kgl. Ges. Wiss. Nachr. Göttingen. Math. Physik.* **2**, 235–257.

Oh, Y.G. [1987] A stability criterion for Hamiltonian systems with symmetry, *J. Geom. Phys.* **4**, 163–182.

Oh, Y.G., N. Sreenath, P.S. Krishnaprasad and J.E. Marsden [1989] The Dynamics of Coupled Planar Rigid Bodies Part 2: Bifurcations, Periodic Solutions, and Chaos, *Dynamics and Diff. Eq'ns.* **1**, 269–298.

Olver, P.J. [1980] On the Hamiltonian structure of evolution equations, *Math. Proc. Camb. Phil. Soc.* **88**, 71–88.

Olver, P.J. [1986] *Applications of Lie Groups to Differential Equations.* Graduate Texts in Mathematics **107**, Springer, Berlin.

Olver, P.J. [1988] Darboux' theorem for Hamiltonian differential operators, *J. Diff. Eqn's* **71**, 10–33.

Otto, M. [1987] A reduction scheme for phase spaces with almost Kähler symmetry regularity results for momentum level sets, *J. of Geom. and Phys.* **4**, 101–118.

Palais, R. [1979] The principle of symmetric criticality, *Comm. Math. Phys.* **69**, 19-30.

Palais, R. [1985] Applications of the symmetric criticality principle in mathemaitcal physics and differential geometry, *Proc. of 1981 Symp. on Diff. Geom. and Diff. Eqn's*, 247-302.

Patrick, G. [1989] The dynamics of two coupled rigid bodies in three space, *Cont. Math. AMS* **97**, 315–336.

Patrick, G. [1990] *Nonlinear Stability of Coupled Rigid Bodies.* PhD Thesis, University of California, Berkeley.

Pauli, W. [1953] On the Hamiltonian structure of non-local field theories, *Il Nuovo Cimento* **X**, 648–667.

Penrose, O. [1960] Electrostatic instabilities of a uniform non-Maxwellian plasma, *Phys. Fluids* **3**, 258–265.

Pierce, A.P., M.A. Dahleh, and H. Rabitz [1988] *Phys. Rev. A.* **37**, 4950.

Pfaffelmoser, K. [1989] *Global Classical Solutions of the Vlasov-Poisson System in Three Dimensions for General Initial Data.* Thesis, University of Munich.

Poincaré, H. [1885] Sur l'équilibre d'une masse fluide animée d'un mouvement de rotation, *Acta. Math.* **7**, 259.

Poincaré, H. [1890] Sur la probléme des trois corps et les équations de la dynamique, *Acta Math.* **13**, 1–271.

Poincaré, H. [1892] Les formes d'équilibre d'une masse fluide en rotation, *Revue Générale des Sciences* **3**, 809–815.

Poincaré, H. [1901] Sur la stabilité de l'équilibre des figures piriformes affectées par une masse fluide en rotation, *Philosophical Transactions A* **198**, 333–373.

Pullin, D.I., and P.G. Saffman [1991] Long time symplectic integration: the example of four-vortex motion, *Proc. Roy. Soc. Lon. A* **432**, 481–494.

Ratiu, T.S. [1982] Euler-Poisson equations on Lie algebras and the N-dimensional heavy rigid body, *Am. J. Math.* **104**, 409–448, 1337.

Reyhanoglu, M. and N.H. McClamroch [1991] Reorientation of space multibody systems maintaining zero angular momentum, *AIAA Conf. on Nav. and Control.*

Rayleigh, J.W.S. [1880] On the stability or instability of certain fluid motions, *Proc. Lond. Math. Soc.* **11**, 57–70.

Lord Rayleigh [1916] On the dynamics of revolving fluids, *Proc. R. Soc. Lond. A* **93**, 148–154.

Reinhall, P.G., T.K. Caughy and D.Q. Storti [1989] Order and chaos in a discrete Duffing oscillator, implications for numerical integration, *Trans. ASME*, **56**, 162–176.

Reyman, A.G. and M.A. Semenov-Tian-Shansky [1990] Group theoretical methods in the theory of integrable systems, *Encyclopedia of Mathematical Sciences*, Springer-Verlag **16**.

Riemann, B. [1860] Untersuchungen über die Bewegung eines flüssigen gleich-artigen Ellipsoides, *Abh. d. Königl. Gesell. der Wiss. zu Göttingen* **9**, 3–36.

Rouhi, A. [1988] The Hamiltonian structure of vortex dipole systems, *preprint*.

Routh, E.J. [1877] *Stability of a Given State of Motion.* Reprinted in Stability of Motion, A.T. Fuller (ed.), Halsted Press, New York, 1975.

Routh, E.J. [1884] *Advanced Rigid Dynamics.* MacMillian and Co., London.

Ruth, R. [1983] A canonical integration techniques, *IEEE Trans. Nucl. Sci.* **30**, 2669–2671.

Salam, F.M.A., J.E. Marsden and P.P. Varaiya [1983] Arnold diffusion in the swing equations of a power system, *IEEE Trans. CAS* **30**, 697–708, **31** 673–688.

Sánchez de Alvarez, G. [1989] Controllability of Poisson control systems with symmetry, *Cont. Math. AMS* **97**, 399–412.

Sanders, J.A. [1982] Melnikov's method and averaging, *Celest. Mech.* **28**, 171–181.

Sanz-Serna, J.M. [1988] Runge-Kutta shemes for Hamiltonian systems, *BIT* **28**, 877–883.

Sanz-Serna, J.M. [1991] Symplectic integrators for Hamiltonian Problems: an overview, *Acta Numerica* (to appear).

Satzer, W.J. [1977] Canonical reduction of mechanical systems invariant under abelian group actions with an application to celestial mechanics, *Ind. Univ. Math. J.* **26**, 951–976.

Scheurle, J. [1989] Chaos in a rapidly forced pendulum equation, *Cont. Math. AMS* **97**, 411–419.

Scheurle, J., J.E. Marsden and P.J. Holmes [1991] Exponentially small estimates for separatrix splittings, *Proc. Conf. Beyond all Orders,* H. Segur and S. Tanveer (eds.), Birkhauser, Boston.

Seliger, R.L. and G.B. Whitham [1968] Variational principles in continuum mechanics, *Proc. Roy. Soc. Lond.* **305**, 1–25.

Shapere, A. [1989] *Gauge Mechanics of Deformable Bodies.* PhD Thesis, Physics, Princeton.

Shapere, A. and F. Wilczeck [1989] Geometry of self-propulsion at low Reynolds number, *J. Fluid Mech.* **198**, 557–585.

Simo, J.C. and M. Doblare [1991], Nonlinear Dynamics of Three-Dimensional Rods: Momentum and Energy Conserving Algorithms, SUDAM Report No.91-3, Stanford University, *Intern. J. Num. Methods in Eng.*, in press

Simo, J.C., D.D. Fox, and M.S. Rifai, [1990], On a stress resultant geometrically exact shell model. Part III: Computational aspects of the nonlinear theory, *Computer Methods in Applied Mech. and Eng.*, **79**, 21-70.

Simo, J.C., D.D Fox and M.S. Rifai [1991], On a stress resultant geometrically exact shell mode. Part VI: Conserving algorithms for nonlinear dynamics, *Intern. J. Num. Meth. in Eng.*, in press

Simo, J.C., D.R. Lewis and J.E. Marsden [1991] Stability of relative equilibria I: The reduced energy momentum method, *Arch. Rat. Mech. Anal.* **115**, 15-59.

Simo, J.C. and J.E. Marsden [1984] On the rotated stress tensor and a material version of the Doyle Ericksen formula, *Arch. Rat. Mech. An.* **86**, 213-231.

Simo, J.C., J.E. Marsden and P.S. Krishnaprasad [1988] The Hamiltonian structure of nonlinear elasticity: The material, spatial, and convective representations of solids, rods, and plates, *Arch. Rat. Mech. An.* **104**, 125-183.

Simo, J.C., T.A. Posbergh and J.E. Marsden [1990] Stability of coupled rigid body and geometrically exact rods: block diagonalization and the energy-momentum method, *Physics Reports* **193**, 280-360.

Simo, J.C., T.A. Posbergh and J.E. Marsden [1991] Stability of relative equilibria II: Three dimensional elasticity, *Arch. Rat. Mech. Anal.* **115**, 61-100.

Simo, J.C., N. Tarnow and K.K. Wong [1991], Exact energy-momentum conserving algorithms and symplectic schemes for nonlinear dynamics, *Computer Methods in Applied Mechanics and Engineering*, in press

Simo, J.C. and K.K. Wong [1989] Unconditionally stable algorithms for the orthogonal group that exactly preserve energy and momentum, *Int. J. Num. Meth. Eng.* **31**, 19-52.

Sjamaar, R. [1990] *Singular Orbit Spaces in Riemannian and Symplectic Geometry.* Thesis, Utrecht.

Sjamaar, R. and E. Lerman [1991] Stratified symplectic spaces and reduction, *Ann. of Math.* (to appear).

Slemrod, M. and J.E. Marsden [1985] Temporal and spatial chaos in a van der Waals fluid due to periodic thermal fluctuations, *Adv. Appl. Math.* **6**, 135–158.

Smale, S. [1967] Differentiable dynamical systems, *Bull. Am. Math. Soc.* **73**, 747–817.

Smale, S. [1970] Topology and mechanics, *Inv. Math.* **10**, 305–331, **11**, 45–64.

Smale, S. [1980] *The Mathematics of Time.* Springer-Verlag.

Sorbie, K.S. and J.N. Murrell [1975] Analytical properties for triatomic molecules from spectroscopic data, *Mol. Phys.* **29**, 1387–1407.

Souriau, J.M. [1970] *Structure des Systemes Dynamiques.* Dunod, Paris.

Sreenath, N., Y.G. Oh, P.S. Krishnaprasad and J.E. Marsden [1988] The dynamics of coupled planar rigid bodies. Part 1: Reduction, equilibria and stability, *Dyn. and Stab. of Systems* **3**, 25–49.

Sternberg, S. [1977] Minimal coupling and the symplectic mechanics of a classical particle in the presence of a Yang–Mills field, *Proc. Nat. Acad. Sci.* **74**, 5253–5254.

Stiefel, E. and C. Scheifele [1971] *Linear and Regular Celestial Mechanics*, Springer-Verlag.

Stofer, D.M. [1987] *Some Geometric and Numerical Methods for Perturbed Integrable Systems.* Thesis, Zurich.

Sudarshan, E.C.G. and N. Mukunda [1974] *Classical Mechanics: A Modern Perspective.* Wiley, New York, 1974; Second Edition, Krieber, Melbourne–Florida, 1983.

Symes, W.W. [1980] Hamiltonian group actions and integrable systems, *Physica D* **1**, 339–374.

Szeri, A.J. and P.J. Holmes [1988] Nonlinear stability of axisymmetric swirling flow, *Phil. Trans. R. Soc. Lon. A* **326**, 327–354.

Tannor, D.J. [1989] Jahn-Teller effects in the photodissociation of ozone. *J. Am. Chem. Assoc.* **111**, 2772–2776.

Tannor, D.J. and Y. Jin[1991] Design of femtosecond pulse sequences to control photochemical products. *Mode Selective Chemistry* (ed. by P. Pullman, J. Tortner, and R.D. Levine), Reidel.

Thurston, W.P. and J.R. Weeks [1986] The mathematics of three-dimensional manifolds, *Scientific American* **251**, 108–120.

Turkington, B. [1989] Vortex rings with swirl: axisymmetric solutions of the Euler equations with nonzero helicity, *SIAM J. Math. An.* **20**.

van der Meer, J.C. [1985] *The Hamiltonian Hopf Bifurcation.* Springer Lecture Notes in Mathematics **1160**.

van der Meer, J.C. [1990] Hamiltonian Hopf bifurcation with symmetry, *Nonlinearity* **3**, 1041–1056.

van der Schaft, A. and D.E. Crouch [1987] Hamiltonian and self-adjoint control systems, *Systems and Cont. Lett.* **8**, 289–295.

van Gils, S.A., M. Krupa and W.F. Langford [1990] Hopf bifurcation with non-semisimple 1:1 resonance, *Nonlinearity* **3**, 825–830.

Walsh, G. [1991] Holonomy and satellite parking, *preprint*.

Wan, Y.H. [1986] The stability of rotating vortex patches, *Comm. Math. Phys.* **107**, 1–20.

Wan, Y.H. [1988] Instability of vortex streets with small cores, *Phys. Lett. A* **127**, 27–32.

Wan, Y.H. [1988] Desingularizations of systems of point vortices, *Physica D* **32**, 277–295.

Wan, Y.H. [1988] Variational principles for Hill's spherical vortex and nearly spherical vortices, *Trans. Am. Math. Soc.* **308**, 299–312.

Wan, Y.H. [1990] Nonlinear stability of stationary spherically symmetric models in stellar dynamics, *Arch. Rat. Mech. An.* **112**, 83–95.

Wan, Y.H. [1991] Codimension two bifurcations of symmeteic sycles in Hamiltonian systems with antisymplectic involutions, *Dyn. and Diff. Eqn's.* (to appear).

Wan, Y.H. and M. Pulvirente [1984] Nonlinear stability of circular vortex patches, *Comm. Math. Phys.* **99**, 435–450.

Wang, L.S., P.S. Krishnaprasad, and J.H. Maddocks [1991] Hamiltonian dynamics of a rigid body in a central gravitaional field, *Cel. Mech. and Dynamic. Astron.* **50**, 349–386.

Weinstein, A. [1977] *Lectures on symplectic manifolds.* CBMS Regional Conf. Ser. in Math. **29**, AMS, Providence, Rhode Island.

Weinstein, A. [1978a] A universal phase space for particles in Yang-Mills fields, *Lett. Math. Phys.* **2**, 417–420.

Weinstein, A. [1978b] Bifurcations and Hamilton's principle, *Math. Zeit.* **159**, 235–248.

Weinstein, A. [1983] Sophus Lie and symplectic geometry, *Expo. Math.* **1**, 95–96.

Weinstein, A. [1984] Stability of Poisson-Hamilton equilibria, *Cont. Math. AMS* **28**, 3–14.

Whittaker, E.T. [1959] *A Treatise on the Analytical Dynamics of Particles and Rigid Bodies.* 4th ed., Cambridge University Press.

Wilczek, F. and A. Shapere [1989] Geometry of self-propulsion at low Reynold's number, Efficiencies of self-propulsion at low Reynold's number, *J. Fluid Mech.* **19**, 557–585, 587–599.

Williamson, J. [1936] On an algebraic problem concerning the normal forms of linear dynamical systems, *Amer. J. Math.* **58**, 141–163.

Wilson, E.B., J.C. Decius and P.C. Cross [1955] *Molecular Vibrations.* McGraw Hill (reprinted by Dover).

Wisdom, J., S.J. Peale and F. Mignard [1984] The chaotic rotation of Hyperion, *Icarus* **58**, 137–152.

Wiggins, S. [1988] *Global bifurcations and chaos.* Springer-Verlag, AMS **73**.

Wong, S.K. [1970] Field and particle equations for the classical Yang-Mills field and particles with isotopic spin, *Il Nuovo Cimento* **LXV**, 689–694.

Xiao, L. and M.E. Kellman [1989] Unified semiclassical dynamics for molecular resonance spectra, *J. Chem. Phys.* **90**, 6086–6097.

Yang, R. and P.S. Krishnaprasad [1990] On the dynamics of floating four bar linkages, *Proc. 28th IEEE Conf. on Decision and Control.*

Zakharov, V.E. [1971] Hamiltonian formalism for hydrodynamic plasma models, *Sov. Phys. JETP* **33**, 927–932.

Zakharov, V.E. [1974] The Hamiltonian formalism for waves in nonlinear media with dispersion, *Izvestia Vuzov, Radiofizika* **17**.

Zakharov, V.E. and L.D. Faddeev [1972] Korteweg-deVries equation: a completely integrable Hamiltonian system, *Funct. Anal. and Appl.* **5**, 280–287.

Zakharov, V.E. and E.A. Kuznetsov [1971] Variational principle and canonical variables in magnetohydrodynamics, *Sov. Phys. Doklady* **15**, 913–914.

Zakharov, V.E. and E.A. Kuznetsov [1974] Three-dimensional solitons, *Sov. Phys. JETP* **39**, 285–286.

Zakharov, V.E. and E.A. Kuznetsov [1984] Hamiltonian formalism for systems of hydrodynamic type, *Math. Phys. Rev.* **4**, 167–220.

Ziglin, S.L. [1980a] Decomposition of separatrices, branching of solutions and nonexistence of an integral in the dynamics of a rigid body, *Trans. Moscow Math. Soc.* **41**, 287.

Ziglin, S.L. [1980b] Nonintegrability of a problem on the motion of four point vortices, *Sov. Math. Dokl.* **21**, 296–299.

Ziglin, S.L. [1981] Branching of solutions and nonexistence of integrals in Hamiltonian systems, *Dokl. Akad. Nauk. SSSR* **257**, 26–29; *Funct. An. Appl.* **16**, 30–41, **17**, 8–23.

Zombro, B. and P. Holmes [1991] Reduction, stability instability and bifurcation in rotationally symmetric Hamiltonian systems, *Dyn. and Stab. of Systems* (to appear).

Zwanziger, J.W., M. Koenig and A. Pines [1990] Berry's phase, *Ann. Rev. Phys. Chem.* **41**, 601–646.

Index